W0082407

The Greater Good

the MODERN SOUTH

series editors
Susan Youngblood Ashmore
& Kari Frederickson

The Greater Good

Media, Family Removal, and TVA Dam Construction in North Alabama

Laura Beth Daws and Susan L. Brinson

The University of Alabama Press
Tuscaloosa

The University of Alabama Press
Tuscaloosa, Alabama 35487-0380
uapress.ua.edu

Copyright © 2019 by the University of Alabama Press
All rights reserved.

Inquiries about reproducing material from this work should be addressed to the University of Alabama Press.

Typeface: Minion and Stone Sans

Cover image: Tennessee Valley Authority (TVA) employee interview-
ing family, 1938; National Archives and Record Administration
Cover design: David Nees

Library of Congress Cataloging-in-Publication Data

Names: Daws, Laura Beth, 1981– author. | Brinson, Susan L., 1958– author.
Title: The greater good : media, family removal, and TVA dam construction in
North Alabama / Laura Beth Daws and Susan L. Brinson.
Other titles: Modern South.
Description: Tuscaloosa : The University of Alabama Press, [2019] |
Series: Modern South | Includes bibliographical references and index.
Identifiers: LCCN 2018029657| ISBN 9780817320089 (cloth) |
ISBN 9780817392215 (ebook)
Subjects: LCSH: Tennessee Valley Authority—Press coverage—Alabama—
History. | Dams—Press coverage—Alabama—History. | New Deal, 1933–1939—
Alabama—History. | Alabama—History.
Classification: LCC TC557.A2 D39 2019 | DDC 627/.809761909043—dc23
LC record available at https://lccn.loc.gov/2018029657

To the north Alabama residents who relocated for Wheeler, Pickwick, and Guntersville Dams

Contents

Tables

Acknowledgments

Our research took us to numerous places across the South, and we are thankful to many talented and knowledgeable archivists and librarians who assisted us along the way. Rebekah Thompson and April Davis at the Limestone County Archives in Athens, Alabama, helped us obtain copies of the *Alabama Courier* (Athens, AL) and *Limestone (AL) Democrat.* Maureen Hill, archivist at the National Archives Southeast Region in Atlanta, provided a great deal of assistance in navigating Record Group 142, particularly with regard to Tennessee Valley Authority (TVA) family removal records and other TVA archives stored there. Pat Bernard Ezzell, Senior Program Manager, Tribal Relations and Corporate History at TVA headquarters in Knoxville, Tennessee, and Nancy Proctor in the TVA library, also in Knoxville, directed us to important, unpublished primary sources about the TVA. Ms. Ezzell's wealth of institutional knowledge about the TVA proved extremely valuable as we compiled this manuscript, and we are thankful for her willingness to share her expertise as well as for her support of this project. Thanks to Joel Walker, organizer of "The Valley of the Dams," a National Archives symposium that gave us the opportunity to deliver our research publicly in September 2014.

We appreciate the patience of and encouragement from Donna Cox Baker and Dan Waterman at the University of Alabama Press. Thanks also to Laurie Varma for her editorial assistance with the final manuscript.

Additional research assistance was provided at both of our home institutions. Elizabeth Cantrell, Joyce Ledbetter, and Barbara Bishop at Auburn University assisted us with finding newspaper articles about the TVA. Mark Gatesman and Aaron Wimer of Southern Polytechnic State University and Kennesaw State University also provided research assistance.

A reviewer of the book proposal suggested finding people who had been relocated by the TVA to include their perspectives in this book, a suggestion that we followed and that improved the research immeasurably. We were able to find these special people thanks largely to north Alabama journalists who wrote articles about our research, asking interested individuals to contact us. We offer special thanks to Kelly Kazek of the *Athens Limestone (AL) News-Courier*, Lee Roop of the *Huntsville Times*, Sam Harvey of the *Guntersville (AL) Advertiser-Gleam*, and the *Arab (AL) Tribune* for taking interest in our project and encouraging their readers to take part.

We were fortunate to talk with several north Alabama residents who were directly affected by the TVA's construction of Wheeler and Guntersville Dams. Maxine Williamson Black, Hazel Moore Thompson, Bill Hardin, George Hodge, Luther Tidwell, Nancy Cabaniss Parker, Bobbie Conner Curry, T. L. Conner, and Paul Conner were children when their families were forced to leave their homes to make way for Wheeler and Guntersville Dams. George Brown was a frequent visitor to his grandmother's farm before she sold it to the TVA, and Eugene Simonson was a child when his family moved to the Guntersville Dam reservoir area for his father's work with the TVA. Their stories were moving and compelling. We are thankful they so generously gave of their time, allowed us into their homes, and so freely shared their experiences with us.

Laura Beth Daws would like to thank:

In 1997, Patti Seibert, who taught tenth-grade history at West Limestone High School in Lester, Alabama, asked me to participate in a team project on the TVA for the state's Alabama History Day competition. It was while researching that project at the University of North Alabama's Collier Library that I first learned the TVA had forced the removal of families while building dams. I'm forever thankful for Mrs. Seibert and her influence on this book project. Thanks also to Dr. Emmett Winn for his support of this work in its earliest form as a paper for his Qualitative Research Methods class at Auburn University, and for his encouragement to keep writing on this topic after my paper was finished. Special thanks to Dr. Mary Helen Brown for providing feedback on an earlier version of this work by serving on my master's thesis committee, for reminding me that you can't edit until you have something on paper first, for her continued friendship, and for her inviting me to present part of this research at Auburn University's Becoming Alabama Conference in 2012. Thanks to Dr. Susan Brinson for serving as my thesis advisor and, eventually, as coauthor on this project. Her support of this work has been incredibly meaningful to me, as has her mentorship, friendship, and encouragement for the past 14 years. Our collaboration on this book has brought me a great deal of professional and personal enjoyment, and I'm thankful to call her a friend.

Since beginning this manuscript, colleagues at three different institutions have provided support for this work. At Georgia Highlands College, Dr. Steve Blankenship and Dr. Jayme Feagin in the Department of History and Mr. Frank Minor in the Department of Humanities often indulged me in conversations about history and the people of north Alabama that helped shape the narrative. Thanks to the Southern Polytechnic State University Department of Digital Writing and Media Arts, as well as the Kennesaw State University School of Communication and Media, for supporting me in my research efforts. My colleagues at all three institutions—who are more like friends—offered considerable emotional and practical support during the writing process, for which I'm especially grateful.

Last but not least, I appreciate the enthusiasm for this project expressed by my family and friends and for their constant cheerleading when I needed it the most. I wish I could name everyone in my circles of friends from Athens, Alabama, the Zeta Eta Chapter of Alpha Delta Pi, my graduate student cohorts from Auburn University and the University of Kentucky, and those I've met through the Georgia Tech tailgate crew. But they know who they are, and they know how much I love them. Lastly, my mother, Carla Daws, my husband, David Milam, and my son, Luke Milam, were especially supportive of and patient with me as I wrote. I'm most thankful for them.

Susan Brinson would like to thank:

I retired during the final stages of writing this book, after 27 years at Auburn University. I was deeply fortunate to have the constant support of colleagues throughout my career, both intellectual and financial. I'm particularly indebted to George Plasketes, Jennifer Adams, and Mike Milford.

I thank Laura Beth for allowing me to participate in the development of this book; her unrelenting passion and interest in this subject has been clear since we started working on her master's thesis in 2004. One of the great privileges of being a teacher and mentor is watching the intellectual gifts of a very few students blossom. Laura Beth is one of those students. I'm honored to call her a collaborator, colleague, and dear friend.

Although I built a career on the ability to communicate, I find myself utterly without adequate words to convey my deep love and gratitude to the dear friends who got me through 2016 and 2017. I refer to them as the protective circle that metaphorically formed around and shielded me. I've learned more about the true nature of love and friendship than I ever knew before. I celebrate Hollie Lavenstein, Liza Mueller, Ric Smith, Frances Kendall, Kim Follin, Laura Beth Daws, Anne Willis, Kathy and John Tamblyn, Mark and Ann Brinson, Marcita Thomas, Peggy Thornton, and Shannon Hodges. A special thank you to Cristiana Shipma, Amelie Marohn, and Erin Slay for their laughter, enthusiasm, and love.

The Greater Good

Introduction

The storied history of north Alabama arguably begins with the Tennessee River, which has served as both a help and a hindrance to the region's development, as both an attractive and a repellent force for the people of the Tennessee Valley. Originally, generations of Native American cultures thrived on the dark loam and red clay soil along the Tennessee.[1] Tribes found that the river and the land alongside it fully provided for their needs. The water made for fertile soil, perfect for growing crops that sustained entire communities. The river provided food in the form of fish and even mussels in the northwest corner of the state. Caves and caverns formed alongside the water served as sacred burial sites for their leaders. Arrowheads, tools, and shards of pottery still lie buried beneath the surface of the land up to the hill country rising above the river, to the extent that tilling up land for backyard gardens today often results in small collections of Native American treasures for landowners. The first inhabitants successfully lived off the land, using the river as their lifeblood for generations before white settlers arrived, pushing them away with the eventual help of the fledgling US government. This would not be the last time Tennessee River area inhabitants were forcibly removed from their land by the government.

North Alabama was a choice destination for settlers, particularly because of the Tennessee River. A 1789 Lexington, Kentucky, newspaper article encouraged readers to journey south promising, "The character of this country is so well established. . . . The soil is as fertile as Kentucky, the climate infinitely more agreeable, better calculated for raising cotton. . . . Its situation is perhaps as convenient as any on the western waters lying on the navigable and beautiful river."[2]

Soon, the influx of settlers resulted in different ways of farming, mostly resulting in misuse and overuse of the land thanks to a dependence on cash crops.[3] Poor farming practices combined with frequent, unpredictable, and uncontrollable flooding meant that, by the 1930s, the land was in generally poor condition despite its potential and location along a steady water source. The river oscillated between a rich resource for the agrarian communities settled on its banks and a source of destruction due to intense floods that could wipe out an entire year's crop in a matter of days.

Though a river generally translates to hearty economic development, the Tennessee's unruly course through north Alabama along with steep variations in channel depth made navigation across the state of Alabama virtually impossible. There were also physical obstructions such as gravel and sand bars and shoals that were legendary for halting trade.[4] Had the river been easier to navigate, the entire area would have benefited from easier access to increased opportunities for business and growth. Its course is unusual, beginning in Knoxville, Tennessee, flowing south out of the Appalachians into Alabama from Bridgeport through Scottsboro, turning northwest from Guntersville until Florence, where it turns southwest, then curves again up to Waterloo and carries on north through Tennessee until it reaches its end at Paducah, Kentucky. While a ship carrying goods from Chattanooga could make it through Guntersville and Decatur, reaching Mississippi and traveling farther north was nearly impossible because of the shallow depth at Muscle Shoals. River vessels simply could not pass through crook-necked areas without risking expensive damage or destruction of their ships. Trade along the river was confined to other routes, excluding much of north Alabama from what might otherwise have been a financially successful cross-country route.

North Alabama's economy suffered like the rest of the country's during the Great Depression, but it had never been an economically thriving area. Years of war, postwar reconstruction failures, and a general lack of industrial development only made conditions worse. Still, thousands of families managed to make a living along the Tennessee. North Alabamians were—and still are—resilient, and their stubbornness and determination kept them on their land, that which their mothers and fathers had forced into productivity, regardless of the problems they faced.

Life was hard for north Alabama farm families in the 1930s, but the people adjusted. They made the best out of what little they had, and they worked toward the goal of a better life for their children. The adults who owned land assumed they would pass down their land, and their way of life, to their children and grandchildren. Of course, many hoped that by the time future generations were old enough to work the land on their own, things would be better, somehow: Educational opportunities would be better, their children would not need to work as hard for as meager results, and the economically depressed South would be pulled out of generations of stagnation. How, exactly, things would get better was a large unknown, and for years, there seemed to be no real solution or help in sight.

That is, of course, until the election of Franklin D. Roosevelt, and the introduction of his New Deal programs. North Alabama families were optimistic when FDR won, promising a New Deal for America, which included pulling the South out of crippling generations-old poverty. In May 1933, Roosevelt signed the Tennessee Valley Authority Act, creating a government agency meant to harness the power of the Tennessee River, improve trade, and bring much-needed cheap electricity to one of the poorest areas of the country. Numerous historians have written of the TVA's importance to the South, especially its importance to north Alabama, which was in particularly dire need of assistance. But in reality, the TVA needed north Alabama's cooperation to be successful. The TVA quickly established a major presence there, taking control of the existing but nonfunctional Wilson Dam, constructing Wheeler and Guntersville Dams, and purchasing land for the Pickwick Dam reservoir area, all within the first 5 years of its existence. For the TVA to be a successful government agency, and to, in fact, save the South, rural north Alabama residents had to do two important things. They needed to (1) immediately start using TVA electricity and (2) serve as a shining example for the rest of the country and world of how successful the TVA's experimental plans truly were. Until the TVA's inception, the government had not entered the utility distribution business, nor had the country embarked on such a far-reaching, comprehensive, regional and social development plan. Millions of dollars were appropriated for construction and other related projects, with little congressional oversight for how that money would be used. While north Alabama stood to gain a new way of life through TVA operations, it was in the government's best interest that north Alabama residents not only accepted but welcomed the TVA into their backyards.

The prospects of an easier way of life, more profits from farms, and non-agricultural employment opportunities offered by the Tennessee Valley Authority understandably generated a great deal of excitement among many residents of the Tennessee Valley. Combined with a deep appreciation for President Roosevelt, most residents were enthusiastic about the TVA's presence in the valley. But, as usually happens, not everyone viewed the TVA positively.

One such person was Hazel Thompson, from Guntersville. Hazel, who was a teenager when the TVA was created, loved the land on which she and her extended family lived. Life for her family was one of hard work, but they were proud of who they were, where they lived, and what they contributed to the agrarian-based economic system in north Alabama. Reflecting on the fact that the TVA required her family to move to make way for the construction of the Guntersville Dam reservoir area, Hazel recalls that her family was unhappy about being forced to leave, despite being paid for their land. The relocation separated her from her grandmother and cousins, with whom she previously had been able to spend considerable amounts of time every day.[5]

Hazel was not the only person who felt this way, nor was she the only person who was facing the prospect of moving yet having nowhere to go. Fortunately for her and her family, the TVA paid them for their land, so they had some money to put toward relocating. Many in north Alabama who were forced to relocate—those who did not own their farms—received no money from the government, yet they were still required to find new places to live. The TVA ultimately relocated nearly 15,000 families living in Tennessee, Alabama, and Kentucky for their dam construction projects. In north Alabama, 2,500 families were forcibly removed from their homes for the construction of Wheeler and Guntersville Dams and the Pickwick Dam reservoir area, which was constructed in nearby Tennessee. What happened to the people living in the backwoods, hills, and "hollers" along the Tennessee River in north Alabama from 1934 to 1938 was disruptive at best and traumatic at worst. Families were forced to leave their homes, and most received very little assistance in finding new places to live. Landowners were paid for their farms and paired with TVA-sponsored real estate agents who helped them find new farms. However, tenant farmers, sharecroppers, and squatters—who did not own their land and who were the majority of families in this area—received no payment and almost no relocation assistance.

The family removal process arguably was the most controversial aspect of the TVA project. Despite the TVA's comprehensive development plans, they largely failed to plan for the social development of the relocated, nor did they provide assistance in the postrelocation phase. Though many supported the TVA, and though the TVA promised to help pull the region out of poverty, many families objected to moving and did so against their will. Yet there were no formal protests against the TVA in north Alabama, and only a few residents spoke out against it even informally.

Why did so many families relocate with minimal protest? The answer to that question is complicated. Some families trusted the government, and FDR, to do what was best for them. Some saw themselves as law-abiding citizens who simply did what they were told. Some felt criticizing or even questioning the perceived big government agency was an exercise in futility, as the poverty-stricken residents had no money or authority that might result in their being able to stay on their land. Some felt their temporary hardships would be worth it, for they might result in a better way of life for future generations. The most important factor, however, that supported TVA operations in north Alabama was their carefully crafted strategic communication plan, designed to shape public opinion and drown out voices of opposition.

This book tells the story of the TVA's communication with north Alabama residents, particularly as it relates to the shaping of public opinion on the Tennessee Valley during the time of construction for Wheeler, Guntersville, and Pickwick Dams. Though much has been written about the TVA, its role in the improvement of social and economic conditions in north Alabama, and the experiences of the relocated families, little attention has been directed to the ways in which the TVA relied on mediated and interpersonal communication to influence the dominant public viewpoint regarding the TVA's presence in the Tennessee Valley. This book illustrates the importance of the TVA Information Office in ensuring that north Alabama residents not only bought into the TVA but spoke positively about it with others in the valley and across the country. We focus on the disconnect between the messages communicated by the TVA Information Office and the experiences of the relocated in north Alabama. In particular, we focus on the importance of newspapers in shaping public opinion about the TVA, pointing out how newspaper stories influenced by Information Office communications left little room for criticism of the TVA or its policies. The historical context presented in the following chapters is designed to help readers understand not just the way of life for Tennessee Valley residents during the Great Depression era, but also the importance and significance of the TVA as a New Deal project designed to help the rural South.

Many primary sources were consulted for this manuscript. We read north Alabama newspapers from 1934–1939 for stories about the TVA. These included the *Florence (AL) Herald* from Lauderdale County, the *Alabama Courier* and *Limestone Democrat* from Limestone County, the *Decatur (AL) Daily* from Morgan County, the *Huntsville Times* from Madison County, and the *Albertville (AL) Herald*, *Boaz (AL) Leader*, and *Guntersville (AL) Advertiser & Democrat* from Marshall County. These newspapers were chosen because their readership resided within the construction areas for Wheeler and Guntersville Dams and the Pickwick reservoir. They were the primary weekly or daily newspapers during the mid-1930s in their respective counties. A search for TVA-related content in each of these newspapers resulted in a set of over 800 articles, editorials, and advertisements that were read for content and tone. Additionally, we read the *Birmingham News*, the *Montgomery Advertiser*, the *Chattanooga Times*, and the *New York Times* to determine state, regional, and national perspectives on the TVA's presence in north Alabama.

Records of the TVA housed at the National Archives Southeast Branch in Atlanta were crucial to this project. We focused on three different areas. First, we explored the Family Removal Case Files for Wheeler, Guntersville, and Pickwick Dams. We read the narratives found on TVA Form 970, a document written for each family who relocated. Each Form 970 contained a cover sheet summarizing demographic data about the family, including the names, ages, health problems, education level, career, and landownership status of each family. Following the cover sheet was anywhere from one to dozens of pages of narratives written about the family and their experience relocating. Narratives offered nuanced aspects of the family's life, such as physical descriptions of the family (both adults and children), housing conditions, directions to the home, their possessions, their income, any problems they faced with relocating, and how many times they were visited by TVA agents assisting with their removal. Additionally, the TVA library in Knoxville, Tennessee, offered a wealth of TVA-published materials such as project descriptions, research reports of the areas in which the dams were to be constructed, the amount TVA paid for land, and postconstruction reports on the impact of the dams on the people and geography of the region.

We reviewed other files within the TVA records group, including those of the TVA Information Office, to determine the scope and extent of the agency's communication plans. Records in this series included annual reports summarizing the productivity of the Information Office, ideas generated by office staff, correspondence between the Director and other TVA officials, and a host of information about the interpersonal and mediated strategies the office employed to distribute TVA-related messages. The records of the Board of Directors dated 1933–1939 were also consulted. They included correspondence about TVA-driven messages, speeches delivered in person and broadcast over the radio, interoffice memos, and letters sent to and received from external audiences.

Finally, we interviewed people who were directly connected with family removal, all of whom were children or teenagers when the TVA came to north Alabama. One interviewee was a child when his father moved his family to Marshall County to work on the TVA's malaria control program. One respondent was a child when his grandmother's farm was purchased by the TVA. Ten additional interviewees were children when their parents' land was purchased by the TVA. They all distinctly remembered the relocation process and what it was like to resettle elsewhere in north Alabama. Interviews ranged in length from 1 hour to 3 hours, were audio recorded, and transcribed by the authors. At the time of the interviews, all relocated respondents were living in north Alabama in close proximity to, if not on, the same land their families purchased after the TVA forced them off their farms.

In addition to exploring how media communicated about the family removal process and shaped public opinion about the TVA, this book explores what happened with the relocated families, as well as the long-term impacts the TVA and its programs had on the Deep South. This is the previously untold story of those who helped facilitate the modernization of the South by giving up land that they owned, rented, or borrowed, trying to make a living through farming, fishing, or any other means possible. This is also the story of the impacts of modernization on a group of people who had much to lose if the TVA experiment had failed. It is the story of poverty, sacrifice, heartbreak, hope, and optimism shared by people who trusted the federal government to lift them out of poverty and give them a fighting chance to catch up with residents in other regions of the country. Most importantly, it is the story of people who sacrificed much—risking their livelihoods for the greater good—so that future generations could live, work, and obtain an education in a modernized South.

1

Life on Depression-Era
North Alabama Farms

The cows always knew when the floods were coming. It meant they had to take a walk across the river. Lonnie Cottingham, his wife, I. D., and their son, Austin, herded the cows up and walked them from their rented land on Knight's Island to higher ground on the Williamson farm, located on the Limestone County side of the Tennessee River across from the island. The Cottinghams, who had lived on Knight's Island since 1898, were all too familiar with this scenario: radio weather reports out of Chattanooga predicted a dangerous flood. Tom Williamson, who owned a nearby 113-acre farm, made his way to the island to help shuttle livestock to the mainland when floodwaters threatened to make their way to this remote area of north Alabama. Usually, cattle, hogs, and chickens lived peacefully alongside their caretakers, who grew corn so green it looked almost black and so tall it looked like trees, especially in the eyes of Tom's 13-year-old daughter, Maxine. Not one for housework, Maxine took every opportunity to work outdoors with her father on the farm, especially if it meant doing a job as exciting as walking the cattle across the water to save them from the potentially catastrophic flood.[1]

Water threatening to rise was one issue. The threat of losing livestock and crops was another. Knight's Island inhabitants did not have much, so what they did possess was incredibly valuable. If they pooled together all their resources, the Cottinghams had 28 hogs, 1 cow, 1 deer, 130 chickens, and 4 mules, all of which could be lost in the flood if they did not take appropriate precautions. These were not pets that needed saving; they were potential food sources. Losing them would mean losing the ability to eat and to earn the meager income they'd grown accustomed to as sharecroppers. Their corn and hay patches were also in trouble, which could mean they would have no money at all at the end of the growing season if their crops died in the floods. If they had planned well, the canned and dried fruit they had as their only other food source would be stored up high enough to stay dry. After securing the animals up hill, all the Cottinghams could do was wait out the flood and assess the damage after it was over.

I. D. Cottingham hated life on Knight's Island, but her husband refused to leave. Three of her four children had moved to other communities across north Alabama, presumably to make a better life for themselves. Though there were few jobs in the area, there were certainly more on the mainland than there were on the island. Lonnie, who worked hard, did the best he could with what little he had for his family of six. His wife, I. D., sensed a better life existed off the island, perhaps even out of the state, but they were seemingly stuck, unable to earn enough money to purchase their own farm elsewhere, yet unable to do jobs more profitable than farming, and unable to see a way off the island.[2]

When they were not preparing for a flood, Knight's Island inhabitants and their north Alabama neighbors along the Tennessee River led simple lives, mostly surviving and rarely profiting, from their jobs as farmers and day laborers. The land on which north Alabamians lived was not *just* land, just as the river was not *just* a source of water, trade, and transportation. For people who had very little in terms of material possessions, one's relationship to the land and the water determined much about one's quality of life. Those who owned, rented, worked on, or squatted on the land along the Tennessee River in the 1930s understood one thing very clearly: the land was their life. The Tennessee River and the land surrounding it both united and divided the families living on it. The river not only influenced but also provided the basis of every aspect of their culture, society, class system, and ways of life. And the river could make or break a year, depending on how kind the floodwaters were to those who lived on its banks.

Much has been written about Depression-era southerners.[3] The 1938 *Report on Economic Conditions of the South* summarizes life in the South best: the "paradox of the South is that while it is blessed by nature with immense wealth, its people as a whole are the poorest in the country."[4] Excellent historical sources address the roots of poverty in the South, particularly at the dawn of New Deal policies intended to help its people out of the shackles of poverty.[5] Lively historical fiction accounts such as *Mud on the Stars, Dunbar's Cove*, and the movie *Wild River* (based on a section of *Mud on the Stars*) provide a dramatized snapshot of what it must have been like for families living on the Tennessee River in the shadow of the Great Depression. But the plight of north Alabama farm families, who were faced with particularly difficult situations due to their location on the Tennessee, bears discussion. The people of north Alabama during the early 1930s suffered from many social and economic problems due to an unprofitable agrarian economy, a general lack of employment opportunities outside farming, and a landownership system that kept sharecropper families indebted to landowners with virtually no chance for upward mobility. Lives were hardened by poverty, poor housing, extremely poor health, social isolation, lack of education, and, in some cases, racism.

Living off the Land in Poverty

Landownership provided the basis for the strict class system present in the valley at that time, one that can be broadly grouped into two categories: those who owned land and those who did not.[6] Generally, land-owning farmers constituted the highest socioeconomic class in the rural Tennessee Valley, simply because they controlled the most valuable commodity: the land. Land, and landownership, also formed the basis of the agrarian economy in north Alabama in the 1930s. Landownership was a sacred resource meant to be passed down from one generation to the next. Landowners had the power to stay in one place and continue taking their chances on crops that could grow on as much land as a family could cultivate.

However, most farmers in the Tennessee River Valley region of north Alabama were not landowners. Farmers in north Alabama during the 1930s operated under a strict hierarchy made up of (from highest to lowest class) landowners, tenant farmers, sharecroppers, and squatters, in a system that contributed to and maintained poverty in the South.[7] Cash tenants pledged their rent even before a crop was harvested. Share tenants owned their livestock and routinely gave one-third of the corn and one-fourth of the cotton they cultivated to their landlord. Sharecroppers did not own livestock or land, but instead worked land owned by someone else in exchange for half the crop.[8] Alabama in particular was home to a disproportionate number of sharecroppers and nomadic squatters who lived wherever they could, two groups of people who found life even more difficult thanks to a system that did not readily permit upward mobility. For example, it was common for sharecroppers to start out their year owing up to a year's earnings to the farmer who owned the land they worked. It was nearly impossible to make a profit in those conditions.[9]

Sharecroppers' and tenant families' lives were fundamentally influenced by the landowner for whom they worked. Many worked essentially as indentured servants under a sharecropping system that seemed impossible to break, as historian Wayne Flynt discusses in *Poor but Proud*. Though they lived on and worked the same land, owners and sharecroppers or renters did not live the same lives. A sharecropping agreement was a financial (though not always fair) agreement. Most sharecroppers did not keep their own financial records, instead relying on the word of the landowner. Even if sharecroppers had been able to prove to a landowner that he was not paying them fairly, it's not likely the tenant would've argued, for fear of being asked to leave the farm, thereby losing their only source of income.[10] Flynt writes that "as one Calhoun County landowner wrote, 'the tenant skins the land and the landlord skins the tenant . . . and he who can skins the landlord.'"[11] Landowners had multiple sources of power. Not only did they own land, they also had the luxury of deciding whether or how to take care of their tenants by continuing to provide land for them to farm. Owners also had the option of removing "bad" tenants who did not want to work and were not producing for them. There were many reasons a farmer might turn out a tenant or sharecropper, but failure to work was perhaps the most common reason. Landowners achieved the highest possible social status, affording them both great power and great responsibility in that they had the ability to take care of others who did not already own land.

There were certainly advantages to owning land. But even Alabama farmers who owned their land struggled more than their counterparts across the country. Most families carried out farm labor without the help of modern machinery crucial to the success of today's farms. In 1930, 13.5 percent of all American farmers had a tractor to help with farm labor, but only 1.7 percent of Alabama farmers did.[12] Farmers in north Alabama worked their crops with plows pulled by mules and however many hands, however big or small, they could find to help. Large farms relied heavily on tenant labor to work them, as well as child labor. Children born into large families grew up in many cases to stay on those same farms, which gradually resulted in less land to work for a growing population.[13]

Children knew better than to complain about the work they had to do on the farm, though, especially children of landowners. At least they had some form of stability. Sharecropping neighbors were often nomadic, roaming from one farm to the next in hopes of a better financial situation for their family. Children were a crucial part of the agrarian economy in north Alabama during the Depression and for a short period afterwards. From the time they were around 5 years old, women like Maxine Williamson Black and Hazel Moore Thompson had been responsible for everyday chores that included milking cows, collecting eggs, feeding chickens, and gathering kindling to keep brooder fires going so the hundreds of chickens living inside it would stay warm. A difficult, and sometimes humorous, chore was holding geese still as another person plucked their feathers, collecting down for feather beds.[14] Picking cotton was par for the course for these families, too. Everyone woke up early on the farm, but when Paul Conner was a teenager, his day started at 3:00 a.m., so he could be the first one at the cotton gin and still make it to school on time. This required hooking up mules and "going to the scales" so his cotton could be weighed early. Getting the mules to help him carry cotton to the gin early in the morning was his responsibility for years, until his father finally purchased a three-quarter-ton Chevrolet pickup truck to help with the farm labor.[15] Luckily, Paul's, Maxine's, and Hazel's status as children of landowners meant they had the opportunity to go to school after their chores were done and the promise of someday inheriting the land on which they toiled. Most north Alabama farm children lacked both of those luxuries.

Life in the Jim Crow South was exceptionally difficult for black farmers, who faced struggles their white counterparts did not. According to one author writing about blacks across the Tennessee Valley, "life on the margins of a poor society required circumspection and docility for many blacks. Protest and organizing against class and race oppression did not enhance a person's chances for survival."[16] Resistance to the laws and cultural norms that further disadvantaged black farmers, then, was futile. According to census data, Alabama's population in 1930 was 64.3 percent white and 35.7 percent black.[17] This ratio largely held true in the nine Alabama counties along the Tennessee River. In 1935, roughly half of all sharecroppers, 42 percent of tenant farmers, and only 16 percent of farm owners in Alabama were black.[18] Black families faced many disadvantages, particularly in terms of landownership, homeownership, and segregated communities. Farms owned by black families were also smaller. While the average white owner's land acreage totaled 125 acres, black owners' farms, on average, were only 75 acres.

Farm income varied depending partially on the forces of nature, over which farmers had no control. The inevitable floods that could wipe out an entire year's crop in a couple of days combined with variable soil quality to make farming a challenge from the outset. Combined with a lack of sustainable farming techniques, farming became a particularly tricky occupation. Despite the constant floods like the one in 1933, people hesitated to leave the riverbanks for the hills to farm. Most farmers sought out the best quality land they could find, which is why land near the Tennessee River was so valuable: It was a farmer's best shot at fertile land. The red clay soil in the hillier country was rocky, and farmers were either unaware of, or unable to implement, newer farming techniques that could coax the land into yielding more crops than the more fertile land along the water. Despite the number of farmers present in the Tennessee Valley during the 1930s, the region suffered from extreme mismanagement of the farmers' most precious resource. Few farmers had the ability to seek out resources and information about how to make the land work for them. As a result, farmers—particularly tenant farmers—applied the strategy of planting as many crops as possible on as much land as they could, without regard for techniques such as appropriate fertilizing, terracing, or working with the natural contours of the land to maximize yield for future years and preserve the soil.[19] This strategy abused the land, robbing it of its potential. Occasionally, optimistic and progressive farmers such as Frank Norwood Conner, educated themselves about cutting-edge farming practices already in use elsewhere around the country including crop rotation and erosion prevention strategies. An eagerness to learn about farming techniques that helped preserve the land and yield better crops was admirable, as was the ability to seek out resources and information about how to make the land work for him. Most farmers simply did the best they could with limited knowledge of farming on whatever land they had, planting the same cash crops in the same locations year after year, participating in the sharecropping or tenant system to maximize land use, a system that resulted in overuse, abuse, and poor profits.

Regardless of the time and hard work farmers devoted to the land, several factors contributed to the inability to break free of the strict class system that revolved around landownership and farming in north Alabama during the 1930s. Varying land quality, lack of machinery to help effectively work crops, lack of proper farming techniques, and, most importantly, farming in a poor economy suffering the effects of the Great Depression all meant that it would be nearly impossible to make a sustainable profit from farming. Still, in the Tennessee Valley, farming was the most viable and common form of employment. If an individual happened to be born into a tenant, sharecropper, or squatter family, was there hope for escaping? In most cases, the answer was no. Renters were rarely able to afford the basic necessities of life and much less able to save enough money to purchase their own farms and establish independence from landowners. This system, which had been in place for generations, needed to be eliminated to give the farmers of north Alabama any hope of a better way of life.

Jobs and Money

According to Bill Hardin, who lived in Hambrick Holler near Guntersville, "everybody [during the 1930s] was either farmers or bootleggers." As for his family? "We farmed, and I guess they did a little bootleggin' too!"[20] The lush, plentiful cornfields, then, served multiple purposes. Who could fault generally law-abiding citizens for doing whatever they could to survive financially during the Great Depression? Many turned to supplementing their income with the manufacture of whiskey from the plentiful corn that grew on the fertile land near the river. Moonshining was illegal, and people were commonly jailed for manufacturing it. But, it did not stop families from pursuing it as both a hobby and supplemental income. And, one would imagine, it also provided a source of entertainment or a way to escape the reality of their situation. Desperate times called for desperate measures.

Among the problems facing Depression-era north Alabamians, poverty stands out as the one common factor, or the one aspect of life that seemed impossible to escape. While the rest of the country reeled from the sting of poverty resulting from the 1929 stock market crash, people in the Tennessee Valley were still coping with the never-ending effects of failed post-Civil War reconstruction efforts. For them, the stock market crash simply meant that hard times would get worse. Maxine Williamson Black's uncle Charlie happened to be in an Athens, Alabama, bank the day of the crash. Maxine, still too young to fully understand what had happened, never forgot what she saw at home later that day, after Charlie made it back to visit with the family. Her extended family scraped together just over $10 they had in cash to hang on to, knowing whatever was in the bank was gone. Her mother cried. And her father paced the floor mumbling about working hard to get through whatever was coming their way.[21]

As Mr. Hardin had noted, most rural families living in north Alabama during the 1930s continued to farm. According to 1930 census data, 33.6 percent of all workers across the United States were involved in farming. But in Alabama, 50.6 percent of the population were farmers.[22] Employment options were limited if one was not involved in farming. Industrial jobs in Alabama in the early 1930s were in the mining, manufacturing, construction, and retail industries, but most of them were located in other parts of the state. Limited industrial opportunities existed in north Alabama, particularly in the cities of Decatur, Huntsville, and Guntersville. Decatur was a railroad town situated on the Tennessee River and is still home to many industrial facilities. Guntersville offered employment at the Saratoga Victory Mills, the Standard Basket Company, the Mountain Hardwood Company, and the Ross-Garden Lumber Company.[23] But for those living in the extreme rural areas in the northwestern and northeastern corners of Alabama, only extremely limited industries or businesses provided employment.

Even if there had been abundant jobs, people living in the area would have found it difficult to keep them due to lack of transportation and burdensome travel time, often several miles away. Most rural residents did not own a car or have reliable transportation. Other viable occupations included fishing, working odd jobs whenever needed, or doing day labor on farms to assist landowners and tenant farmers. For those unable to work, such as the elderly or disabled, the best hope was being taken in by a relative or neighbor. All most residents could do was farm.

Ironically, despite the number of farmers and prevalence of farmland, farming was not profitable enough to sustain most families above mere subsistence. Farming was a popular, but not necessarily profitable, occupation. The overabundance of farmers and farms also contributed to poor economic conditions; too many farmers resulted in a saturation of the cotton market, which drove down the price of the crop and resulted in farmers struggling to grow enough to break even. An estimated 207,000 cotton farms were located in Alabama in 1930; 70 percent of those farms were managed by tenant farmers.[24] While half the population of Alabama workers were farmers, they only brought in 21.9 percent of the state's total income. Annual per capita income for nonfarmers was $428.16, but only $170.25 for farmers.[25] Even landowners did not make large profits. Most renters did not make enough money each year to save even a little of their earnings. And, most farm labor was seasonal, especially for cotton pickers and choppers, so income was inherently unsteady.[26] Like most class systems, one was born into a class that was almost impossible to escape. Even the highest class in rural north Alabama during the 1930s did not have much compared to the upper socioeconomic classes of their neighbors in other parts of the state or country. As a region, the South was generally poorer than other areas of the country. The most prosperous state in the South still ranked lowest in per capita income of any state in any other region of the country.[27] In 1929, Alabama was ranked 26th in the United States in terms of total income, and Alabama's per capita income was 45th out of 48 states.[28] Though per capita income for all of Alabama at this time was $330, it was much lower along the Tennessee River. The effects of the Great Depression can be seen in the drop of per capita income among all counties located along the Tennessee River. Table 1 summarizes the percent change in per capita income from 1929 to 1935.

In 1929, the average income for farm families in northwest Alabama was $852 per year, while families in northeast Alabama averaged an income of $776 per year.[29] That translated to an average monthly income of $68.75 in the northwest and $64.66 in the northeast. This was well below the national average for farm income in the United States during that time, which held at roughly $100 per month.[30] Money was even tighter for tenant families and sharecroppers, who might make much less than the average per capita in each county. One report suggests that sharecroppers across the South made between $38–87 annually, which translated to as little as 10¢ each day.[31]

For north Alabamians in the 1930s, the Great Depression largely meant

Table 1. Percentage of change in per capita farm income among counties in TVA reservoir areas, 1929–1935

County	Per Capita income, farms (1935)	Percent Change, 1929–1935
Colbert	102	-45.7
Jackson	85	-33.6
Lauderdale	94	-46.3
Lawrence	98	-40.2
Limestone	101	-43.3
Madison	111	-31.5
Marshall	112	-38.5
Morgan	97	-40.1

Source: Adamson, *Income in Counties of Alabama, 1929–1935* (Tuscaloosa: Bureau of Business Research, School of Commerce and Business Administration, University of Alabama, 1939), 69

an extension of the hard times to which they had grown accustomed, if not fully accepted. Resignation to their fate as poor and an acceptance of their lot in life runs throughout many of the stories of those who were living along the Tennessee River during that time. These people were not necessarily attempting to farm for a profit but instead to do their best to survive until the next year. Clearly, poverty levels in the South were problematic, and much national attention was directed on the South even before the stock market crash. Although first published in 1941, James Agee and Walker Evans's *Let Us Now Praise Famous Men* reports the lives of tenant farmers in west central Alabama during the summer of 1936. Government attempts at relief in north Alabama provided a steady but very small income—$13.50 per month, which often had to support entire families with multiple children—and were insufficient.[32] Even so, this stipend was higher than the average relief of $8 per month across the state.[33] Some families expected governmental assistance but did not receive it, because either they did not understand the process and were unsuccessful or they never applied for it.[34] According to 1930 census data, 17.2 percent of all Alabamians were on some form of relief.[35] This translated to over 98,000 families, 19 percent of all black families and 16 percent of all white families, receiving some form of federal relief.

Rural farmers in north Alabama lived in deep financial distress and had lived that way for many years. Extended poverty breeds a certain feeling of helplessness, as poverty in this case did not simply mean a lack of money or resources. For people who never had much in the first place, their daily lives concentrated endlessly on repairing decrepit shacks and scrabbling together what they could for their next meal. A better life with more money and resources seemed like an unattainable dream. It is difficult, if not impossible, to discuss the myriad aspects of poverty as separate; health, housing, financial, and educational problems all stemmed from a general lack of resources, and they all fed off each other contributing to a vicious cycle that seemingly had no end.

Housing

Not everyone lived on land in north Alabama. Some found creative housing options, like the Preston family, who lived in a previously sunken houseboat tied up on the riverfront in the Morgan County seat of Decatur.[36] Emmett, Virginia, and their children Lillie Sue and Dimple inhabited the houseboat, which kept them in close proximity to the river, where Emmett fished for a living. Houseboats were not uncommon abodes for families along the river, especially for nonlandowners who opted to fish for a living as opposed to eking out an existence as tenant farmers. Like their nomadic land dwellers, residents of houseboats would simply move on whenever they exhausted their opportunity in one area of the river.

The places in which farm families lived ran the gamut from well-constructed to decrepit: some were adequate, most were too crowded, and others had all but fallen down. Homes in the poorest areas of the Tennessee Valley were not the sacred places of rest, relaxation, and personal expression our homes are today. They were crude structures that provided just enough shelter to keep out rain and wind. According to the 1930 census, the median value of owned farm homes was $1,135. In Alabama, however, the median value was less than $500, which was the same value of tenant homes in both the United States and Alabama.[37] Alabamians as a whole did not own as many homes as did their neighbors across the country. Table 2 illustrates the disparity among homeownership for Alabamians, particularly in north Alabama.

It was even more difficult for black families to achieve the status of homeowner, not just in Alabama but across the country. Black homeowners constituted only 19.3 percent of all farm homeowners, but 77.4 percent of all farm home rentals.[38]

Frame, box-type houses with a minimal number of rooms were the most common type of dwelling in the reservoir areas, but some lived in worse conditions that should have been viewed as unfit for habitation. Like most areas of life, one's landownership status influenced housing. Landowners who knew they would be in the same place for many years had more incen-

Table 2. Percentage of people who owned or rented homes, 1930

	Owned (percent)	Rented (percent)	Unknown (percent)
United States	46.0	51.0	1.9
Alabama	33.5	64.5	2
Wheeler Dam reservoir area	7.0	93.0	—
Pickwick reservoir area	25.0	75.0	—

Source: US Census Bureau, 1930 census. Vol. 6, *Population & Families*, Tables 2 and 41. TVA Office of the General Manager, Information Office Correspondence Files, 1933–1946, Box 23.

TVA Office of the General Manager, Information Office Correspondence Files, 1933–1946, "TVA Housing Development at Pickwick Dam," RG 142, Box 23, National Archives–SERA.

tive to put effort into the upkeep of their generally simple homes. Very few farm homes were painted on the outside; a painted home meant the family had achieved the highest status in homeownership. Some of the nicer farm homes had adequate furnishings and porches, such as Hazel Moore Thompson's grandmother's home. Hallie Entrekin lived in a well-kept home with not one but two porches. Hazel, her parents, and her four brothers and sisters lived next door in a five-room house of similar construction and quality.[39] Both homes had meager furnishings such as beds and a table with chairs.

The Cottinghams of Knight's Island, the same family that routinely moved cows across the nearly dry river bed when the floods came, lived in a more common type of house for the backwoods of north Alabama. Their home had a leaky roof, bad floors, and a crumbling foundation.[40] Though they were tenants, they had occupied the same home for several years, with little money or incentive to make basic, needed repairs. Homes for share-croppers, tenant farmers, or squatters usually were not as stable as land-owners' housing and were not intended to be permanent structures built on sturdy foundations. The Stone family of Guntersville lived in a home representative of the typical sharecropper house: it had been occupied by many different families over the course of several years, and as a result was a run-down, two-room shack in a poor state of repair.[41] It was, however, livable and provided basic shelter from the elements.

It was common for large families, and in some cases, multiple families, to live in a house or shack under extremely cramped conditions. This was particularly true for the nomadic squatter class. Overcrowded homes were not a unique issue facing farm families; towns across the South were overcrowded.[42] Some families resorted to occupying any structure they could find, for as long as they could hold on to it. One Marshall County extended family lived in a one-room shack that was previously a store building, with no less than eight people inhabiting it at any given time.[43] Such families lived in destitute circumstances, sometimes because of issues even farther beyond their control than the class system into which they were born. When his brother was killed, leaving behind a wife and five small children, Lessie Green of Limestone County took the family in to his humble two-room shack, furnished only with one bed.[44] Almost inconceivably, some families were forced to live with animals not meant for domestication. Sixteen members of two unrelated Jackson County, Alabama, families shared a three-room house, one that had cracks in the wall as large as windows, with the chickens they owned—which were their sole possessions.[45] Some structures were not "fit to keep livestock in, let alone human beings,"[46] "made of scrap iron and junk; positively not fit for human habitation;"[47] and "a miserable hole."[48] Some families, resigned to their fate, were "content to live in [a] shack in rags without the bare necessities of life."[49] And there were some families who survived in rudimentary tents, which provided minimal shelter and safety.[50] Regardless of type, most homes did not have electricity or running water, problems that contributed to many of the health and sanitation issues facing rural southerners.

Sanitation and Health

Whenever he found himself able to spare a ham, no doubt taken from one of the hogs he owned and slaughtered, Thomas Williamson made it a point to take one to Dr. Johnny Crutcher, a family physician whose office was in nearby Athens. Thomas's wife made it a point to bring eggs or even chickens to his office if she happened to be in town. So, when Maxine, their youngest daughter, came down with typhoid fever on the Fourth of July, 1933, with little chance to live, Dr. Crutcher did perhaps more than he would for other patients. He arranged for one of his colleagues, who had just earned her nursing degree, to stay with the Williamsons for a month and nurse Maxine back to health. The Williamsons certainly did not have the money to pay for that sort of long-term, intense care, but Dr. Crutcher took care of his own in whatever way he could. Though they were not related, the Williamsons thought of Dr. Crutcher as family; he'd taken such good care of their whole family their entire lives. Maxine's experience recovering from typhoid fever was an anomaly. She understood how lucky she was to have access to medical care that many others did not. In the rural south during the Great Depression, bartering for goods and services was common, as was the kinship felt among friends who became family. But Maxine's good fortune in this situation stemmed from her status as a landowner's child, which gave her family personal access to a doctor at a time when many did not have the ability to seek treatment for less severe, but just as threatening, illnesses.

There were other benevolent doctors across north Alabama who did what they could to help those who needed it even when families could not afford to pay for medical care. Dr. W. R. Rousseau of Lauderdale County treated a number of his neighbors for a variety of illnesses. One patient in particular, widower Thomas Green of Lauderdale County, had health problems that made it difficult for him to take care of himself and his seven children. Dr. Rousseau was committed to helping him through the pneumonia he developed in the fall of 1935, even though he knew Green could not afford to pay him for treatment. Nor did Green have as much of a chance of surviving given his living conditions: a debris-scattered yard and a dirty house—complete with slop for the hogs in the kitchen—that kept getting dirtier with every day Mr. Green got sicker.[51] His daughter, Aline, who had acted as the mother figure to her six siblings since their mother's death, was also sick that fall. She'd spent 8 weeks in bed, and as a result, the entire house suffered from lack of food and lack of attention. There was only so much Dr. Rousseau could do for them, but, defying the odds against them, both Thomas and Aline recovered.

Others, like the Rice family in nearby Madison County, were not as fortunate as the Williamsons with nursing their sick children back to health or with the stroke of luck the Green family seemed to have with their health. Indeed, they were luckier than some, as most of their 10 children survived past toddlerhood. Their baby, however, died of severe diarrhea.[52] It was not uncommon for rural tenant families to lose children to what are today routine and minor illnesses. Malaria was rampant in the area and would strike entire families.[53] Poor drainage resulted in large areas of standing water that attracted mosquitos carrying the preventable illness.[54] Each year across the South, more than 2 million men, women, and children were infected with malaria.[55] Vaccines were available for some illnesses, like malaria, but not everyone had equal access to them. Generally, the rural farm families of north Alabama had very poor health care, worsened by deteriorating housing conditions and a host of sanitation issues facing residents.

The lowest-income sections of the South have been described as "a belt of sickness, misery, and unnecessary death."[56] During the Great Depression, a lack of general resources, accessibility to doctors, and money for medicine meant that illnesses routinely treated with affordable antibiotics today could be deadly. Families had little access to running water, and even accessing clean water was challenging on occasion. Nationwide, running water was available to 15.8 percent of households, but only 2 percent of Alabama residents had it. Lack of running water translated into a lack of indoor toilet facilities: 8.4 percent of Americans had indoor toilets, but only 1.2 percent of Alabamians did. One report from the era noted that no farm homes in the extreme northwest corner of Alabama had an inside toilet or a bathtub.[57] This issue resulted in even more health problems; contaminated surface wells and a lack of proper sewer systems contributed to the prevalence of typhoid fever, hookworm, and anemia.[58]

County sanitation and health departments did what they could, but most simply lacked the funds needed to keep residents healthy. For example, the Marshall County Health Department in northeast Alabama had an annual budget of $6,800, which was used to pay a four-person staff and operate an entire clinic.[59] A number of public health issues proved impossible to solve without also solving the problem of substandard living conditions. Because housing was of such poor quality, if one fell ill, it was difficult if not impossible for the sick to find rest in a clean, comfortable, healthy environment.

These problems, much like the agrarian-based class system, needed more intervention than local, state, and federal relief agencies were able to provide. Even with the assistance of local doctors who would assist those in need pro bono, it was impossible to effectively address the rampant malaria and lack of access to good health care in north Alabama given the lack of unified, dedicated, federal attention to the problem and given the social isolation many of these residents faced. Furthermore, living in geographically isolated areas not only made access to doctors limited, but it also contributed greatly to the poor health and poor living conditions for residents.

Lack of Food, Resources, and Education

It was impossible to get to Clarence Tidwell's house by car. As residents of Honeycomb Holler in northwest Alabama, near Guntersville, the Tidwells lived about a mile off US Highway 241. The best way to get to their house, if you had any business visiting, was to park at Mr. Cooper's home, which was about a quarter mile past John Whitaker's store, and walk across the fields to access the home.[60] This was all, of course, assuming the weather was cooperative and the river was agreeable. During the winter, there was no real way to access their house—or for the Tidwells to leave—save by mule across a mountain. Nearby Hamrick Holler, where neighbor Bill Hardin lived, was equally as isolated. There was only one way into, and one way out of, the Hardin homestead—around the base of the mountain.

Isolation, sanitation, health, and food issues went hand in hand for the rural poor of north Alabama. Paved roads led to gravel roads, which led to wagon roads, which led to dirt roads, which led to paths that helped the occasional outsider navigate the hills and hollers along the Tennessee River in the 1930s. This seclusion was sometimes by choice: some wanted to be away from the rest of civilization, and others wanted to hide illegal boot-legging activity. But isolation also contributed to the overall poverty in the region and sometimes also led to fear of outsiders. For example, one north Alabama family, the Durhams, reported that their children were "afraid of strangers" because they lived "in such a remote section and [had] no contact with other children."[61] Remoteness meant difficulty in traveling to schools, churches, stores, towns, and other places to congregate with others. Some who lived on small islands on the river, such as Knight's Island, were further isolated when floods ravaged the area, as many of the islands were rendered inaccessible. This isolation also contributed to one of the great ironies of the Tennessee Valley in the 1930s: Despite the number of farmers, there were great food shortages. Isolation meant that farmers were unable to access what little surplus their neighbors might have had, and families were, therefore, forced to fend for themselves.

Tenants and day laborers struggled to cultivate subsistence gardens due to a lack of available land, poor land quality, or the ability to stay in one place long enough to reap a harvest. Bartering or trading was also not an option for families who had few or no possessions. One tenant family living on the land owned by Paul Conner's father ventured out to a peanut patch after Paul's family harvested the crop. There were inevitably a few peanuts left behind, mostly the tiny ones that had not developed properly. This par-ticular tenant family of four had scoured the field for those leftovers, parch-ing them on a small tin stove and eating them while sitting on the floor of their two-room shack with nothing inside but a mattress. Seeing that family eating the burned peanuts struck the young Paul as heartbreaking. But food scarcity was a common problem among sharecroppers. In fact, food was so scarce that north Alabama poor frequently ate clay and dirt, a condition known as pica.[62] The red Alabama clay contained more nutrients than some of the food available, and for the poorest residents who had no subsistence garden, no way to barter for food, and no money to buy food staples, the dirt was sometimes the best thing they could eat.

Tenant families survived on simple foods such as cornbread. Two fam-ilies in the Guntersville area had survived on cornbread and milk, or only cornbread, for some time. Their situation was one of dire need, yet they had no real assistance to solve their problem.[63] Out of sheer desperation, some tenant families resorted to theft. One man stole a pig from a neighbor, likely understanding he'd get caught, simply because he could not see another way to feed his family.[64] Access to clean water was an issue for some, particularly in the most destitute areas of northwest Alabama, and access could be especially problematic if a tenant family lived on a stingy owner's land. One owner refused to allow his tenants to retrieve fresh drinking water from a nearby well, meaning the tenant family was forced to drink dirty creek water.[65]

More benevolent landowners, however, who had enough food to spare and put in the hard work of preparing and storing food for times of shortage, shared their food with those helping them work their land. Sharecroppers helping Mr. Williamson work his farm in Limestone County were treated every day to lunch and dinner, prepared by Maxine's mother, containing a variety of meats and vegetables grown in their garden. The smokehouse and numerous hogs they owned translated to no shortage of food for themselves and those who were lucky enough to work with them. Those who could diligently canned fruits and vegetables, keeping stockpiles for hard times. Bill Hardin's family was similarly lucky: They "grew everything [they] ate." They took their corn to the grist mill and sold eggs or butter to earn money for staples like sugar, flour, salt, and pepper. Such trades were facilitated by the "peddler truck" that brought groceries to those living in the farthest-outlying areas of north Alabama.[66] Even if the peddler truck made its way to a community, it did not guarantee that all families would enjoy the riches it brought along with it. One's socioeconomic status as landowner, renter, or squatter directly influenced one's access to nutritious food. Food scarcity was worse for renters or squatters, who would forego subsistence gardens in an effort to maximize crop profits and thus had nothing to barter.

Access to Electricity

Rural folks paid for their isolation in a number of ways, including lack of access to utilities their city-dwelling neighbors had. Paul Conner's father was no exception. He was fully accustomed to living on the farm without modern conveniences, but when he had the chance to indulge, he took it. To him, indulgences did not translate to buying a fancy item of clothing or enjoying a meal at a restaurant. Frank Norwood Conner's indulgence was ice. He loved ice more than just about anything, but because his farm had no electricity and no refrigeration or freezing capabilities, he rarely had it. On Saturdays, Frank took his family to the closest store, where he could buy a large block of ice. He used it to make ice cream for his family as a Saturday night dessert, in an old 6-quart ice cream freezer. His children ate the sweet treat until there was nothing left. They generally kept the block of ice in a makeshift cooler, so it may have lasted until Tuesday if they were lucky. The entire family—but particularly Frank—would wistfully watch as the last sliver of ice dissolved in their tea at lunchtime. They knew it would be at least 4 more days before they would have the chance to buy ice again, and even then, only if there was any leftover cash.

The seemingly smallest conveniences many take for granted today, such as keeping milk cold, was a challenge for those without electricity. Beryl Tidwell's parents and neighbors developed a system that worked well within their community. A cave down the road—still technically within their community, but quite far from where most lived—had a spring that stayed cold even in the hot summer months. Neighbors from all around stored their milk there. No one ever bothered anyone else's milk, and no one doubted their neighbors honored the system.[67]

Life on north Alabama farms was made more difficult due to a lack of modern conveniences, such as electricity. Daily household chores including laundry, housecleaning, and cooking were much more challenging due to lack of electricity and other modern conveniences. Prolonging food life without electricity was particularly challenging; refrigerators and freezers were essentially nonexistent in rural areas of north Alabama. Not only were refrigerators too expensive for most families; there was no electricity with which to keep them running.

In the years leading up to the Great Depression, private utility companies, such as Alabama Power based out of Birmingham, had been working to bring electricity to the most outlying rural areas of the South. However, their efforts were slow, and electricity was relatively expensive. For those who couldn't afford clothes or food, electricity was a privilege they could not afford, no matter its benefits. The 1930 census reported that 13.4 percent of Americans had electricity, compared to 12.5 percent of Alabamians. The closest rural residents got to having electricity was a Delco or battery-operated system that provided minimal electricity to a small farm house. The more well-to-do landowners, such as the family of Maxine Williamson Black, were generally the ones with a Delco system. This systemic problem was clearly not being solved fast enough by the private utility companies, whose main interest was profits, not modernization.

Not only was Alabama lagging in access to electricity; the state had not yet caught up with the rest of the country in terms of other modern conveniences. According to the 1930 census, 34 percent of Americans had telephones, while only 7.5 percent of Alabama farmers did.

Most rural farmers in north Alabama had little cash, and, predictably, few material possessions. For some families, their only possessions would have been meager house furnishings, such as a chair or a bed for an entire family, that amounted to less than $20 in value. Some owned just one or two farm animals in addition to their small homes. And some, particularly those who lived in tents, had nothing but makeshift shelter to keep themselves covered. Living in these conditions—with inadequate housing, no electricity, no money, and no resources—for many, resulted in an attitude of hopelessness, a belief that attempts to escape the system that entrapped them would be fruitless, and a strong sense of desperation for a better life.[68]

Desperation manifested itself in many ways. In Guntersville, the Tennessee River ferry cost just a nickel per trip, but some who needed to cross the river did not have a nickel to spare. Some folks resorted to carving out a hole in a cotton wagon and hiding multiple people inside, hoping to sneak people across the river without paying for them.[69] The elderly who were no longer physically able to meet the demands of farming, particularly elderly women who lost husbands (and children), suffered disproportionately. Relief for them was often less than it was for entire families; "old age relief" of $4 a month generally was insufficient to cover rent and food. One resident tried to help herself by washing others' clothes, when she was able, but she eventually resorted to moving in with her son and his family in an already-crowded home.[70] And visitors to one particularly destitute home in the Lim Rock community of Jackson County would have found it difficult to tolerate: The home had no ceiling, and there were large gaps in the thrown-together walls. It was not clean; junk and debris filled the yard while the interior was putrid, swarming with flies. The family shared their home with chickens, who occasionally took up residence on the meager furnishings: a table, makeshift beds, and chairs. The family could not afford toys for any of their six children, so the baby had resorted to playing with a lizard. He kept playing with it even after an older child crushed it with a pan. Older kids had to make their own entertainment, too, after all.[71] There were as many similar stories as there were poverty-stricken families across north Alabama.

Education

Farm families used whatever they did have to make the best life possible for themselves. Take, for example, Hot Water, who was Maxine Williamson Black's horse. She served dual purposes. Maxine's father occasionally needed Hot Water in the field for plowing. But most of the time, Maxine and her brother could ride her the 3.5 miles it took to get to the nearest school. Hot Water was a reliable source of transportation for several years, and one of the reasons Maxine was able to continue her early education. Of course, not everyone was fortunate enough to have their own horse. Some had to walk the multiple miles to school from home. Trips to school ranged from adventurous to potentially dangerous. Hazel Thompson's father would carry her and her siblings to school in a boat when backwaters rose at their home. During those times, a boat was the only way to escape the area. Her father, who could read but only had a third-grade education, was determined that his children would go farther in school than he had.

Alabamians in the Great Depression era were "uneducated but not unskilled or unintelligent."[72] Still, children of landowners, with higher social status, material wealth, and financial resources, found it easier than others to keep up with their education. Sharecropping families had a harder time, particularly those who came from generations of uneducated people. James Torance was the only one of his parents' six children to survive to adulthood. He'd lived a hard life of tenant farming and found himself married with nine children (eight of whom had survived past age 2), living off Spring Creek in Lawrence County, Alabama. He and his wife, Ellen, were exceptionally hard workers, trying their best to make a better way of life for themselves and their children. But neither of them ever learned to read, and they were saddened by the limited opportunities they could provide for their family. Like many other families, the Torances understood that education was one way out of the sharecropper system. Unfortunately, for them and for many other families, circumstances made attending school nearly impossible. The Torances lived 2 miles from the nearest gravel road, had no reliable transportation, and no way to get their children—who were starving—to school every day. For at least 2 months, they lived on butter, bread, milk, and whatever leftover corn they could find after the crop had been harvested. A similar educational situation faced Bill Hardin. The Hamrick Holler resident had very few educational opportunities due to the geographic isolation of his community.[73] Bill, however, was determined to finish high school, and as a child of landowners, he was able to do that. He earned his high school diploma from a small schoolhouse in the country before moving on to successful military and civilian careers.

Landowners' children generally received educations because, although the children had farm chores, the majority of the farm labor was provided by tenant farmers, sharecroppers, and day laborers. This is yet another example of how one's landownership status influenced every area of life: Children of landowners had more educational opportunities than children whose families worked on rented or shared land. For those families who moved frequently and did not have a regular tenure with a landowner, it was even harder for their children to attend school regularly. A strong sense of determination and hard work—personality traits that can be applied to so many of north Alabama's poor during the 1930s—likely kept some children in school, despite all odds and in the face of many challenges. Regardless of one's landownership status, most parents deeply desired the opportunity for their children to earn an education. Some pushed their children to attend school at very high costs, like Paul Conner, who started every day before dawn so he could complete his work on the farm and still make it to school on time.

Geographic isolation, lack of reliable transportation, and the necessity of working to support families made it difficult, if not impossible, for many living in the Tennessee Valley during the 1930s to receive an education. Part of the problem was a financial one, stemming from issues with the state of Alabama's lack of ability to spend adequate money on educating its youth. At the time, "the per pupil investment in the State of Alabama . . . is $89 as compared with $224 for the United States as a whole. The State of Iowa, in which the rural–urban balance is similar to that of Alabama, has invested $222 per pupil."[74] The disparity in the funds spent per student did not stem from the state not wanting to educate its residents. The South, as a whole, simply could not compete with the northeastern states, where average property taxes per person totaled $1,370 compared with $463 per person in the South.[75] Still, the South as a whole spent proportionally more of their tax income on education.[76] With so little funding available for Alabama children's education, it was extremely difficult to educate the state. But greater economic factors influenced the problem of adequate education in the South as well, particularly for Alabama residents who were disadvantaged educationally in the region. In 1930, school-aged children across the country attended 143 days of school per year; in Alabama, students attended just 114 days on average.[77] As one might expect, illiteracy was one problem stemming from the lack of educational opportunities in the valley. Nearly 13 percent of Alabama residents were illiterate in 1930, as compared to 4.3 percent of all Americans.[78] And illiteracy was even more of a problem in the north Alabama counties along the Tennessee River, where rates of illiteracy reached 16 percent. Table 3 shows how in all counties across north Alabama, illiteracy was a major problem impacting the population.

It was virtually impossible for black children to get an education on par with their white neighbors. Considering the inadequacy of education that already existed for white students, black students had even fewer opportunities to become educated on the level of their counterparts throughout the country. In the state of Alabama, leaders blamed a lack of overall funding for this disparity: "The expense for education is high in proportion to the income of the people. The present income is not great enough to permit the state, or the county, to provide adequately for the education of all the children of both races."[79] Across north Alabama during the early 1930s, 82.5

Table 3. Illiteracy rates in counties bordering the Tennessee River, 1930

County	Illiteracy Rate (percent)
Colbert	16.6
Franklin	7.0
Jackson	11.5
Lauderdale	8.6
Lawrence	12.8
Limestone	16.0
Madison	11.9
Marshall	8.1
Morgan	9.6

percent of black children were educated at a level below where they should have been based on age; 78.4 percent of white children fit that description. Facilities for black schools were also inadequate. In the mid-1930s, the town of Guntersville had both elementary and high schools for whites, but only "two Negro grade schools . . . housed in Negro churches."[80] The black schools were "seriously crowded."

Conclusion

A visitor to north Alabama during the early 1930s would find lives of almost unimaginable destitution. Certainly, many land-owning Alabamians lived modestly, with plentiful food, running water, access to health care, and stable communities. But the strict social and economic structures in the region resulted in large numbers of sharecroppers, renters, tenant farmers, and squatters whose lives varied in degree from want to destitution. Federal and state relief efforts were not meeting the dire needs of Alabama's poorest residents.[81]

People living along the Tennessee River in north Alabama during the Great Depression had access to relatively few resources and almost no opportunity to free themselves from a social class system and culture that maintained a life of poverty. Rural north Alabama had not yet caught up to the rest of the country in nearly every aspect of modern society. The countryside had not yet received electricity, there was little running water, more people than the national average were illiterate or uneducated, and many were socially isolated with few outside influences.[82]

Such influences might have been helpful to a culture that was barely scraping by. Most were aware that more help than existing agencies were providing was needed. According to one government report, the area needed a "sound agricultural–industrial program . . . which will lead to the profitable utilization of natural resources . . . [and] permit the people to raise their standard of living by increasing their individual incomes."[83] Who would create or fund that program was yet to be determined.

Despite all the socioeconomic problems facing north Alabamians in the 1930s, residents maintained a "stamina and determined optimism" that helped them during tough times.[84] They needed, however, a substantial amount of help to escape poverty and modernize the outdated farming techniques that were stripping the land of valuable mineral resources. Federal relief was failing them. There was not enough money coming into the area to pull the families out of poverty.

Few would disagree that the area needed intense help. But little did these farmers in the hills and hollers of north Alabama realize that they would soon be asked to bear a disproportionate burden of the change that was quickly coming to the area thanks to a promise from one of the country's most beloved political leaders.

2

The New Deal and the TVA Plan

Poverty in the South was staggering and intransigent. The lack of rebuilding during the Reconstruction era following the Civil War, systemic racism, a tenant-based farming system, destructive and barely effective agricultural practices, and a river that routinely flooded meant that many in the South were already suffering grievously when the Great Depression started in 1929. Elected in 1932, President Franklin D. Roosevelt was sympathetic to the problems and needs of southerners and, during his presidential campaign, offered a "new deal" by promising federal government assistance to them.[1] Many southerners were willing to try Roosevelt's plan. As one historian put it, "Roosevelt's election brought hope to southerners [who] knew little about what the 'new deal' held for them. But after twelve years of Republican presidents and four years of Depression, southerners looked to the future with renewed optimism."[2] The federal government's goal proved daunting.

Bolstered by virtually unbounded optimism, twenty-seven New Deal agencies swept through the South with good intentions, social theories, and federal dollars.[3] Recognizing the vast problems of poverty in the Deep South, the federal government implemented many relief and rehabilitation efforts in the early to mid-1930s. Notable among the agencies, in addition to the TVA, were the Farm Security Administration (FSA) and the Federal Emergency Relief Administration (FERA), which lent money to hundreds of thousands of farmers, especially those in the Deep South. While the money, mostly in the form of subsistence grants, offered short-term help to farmers, it was often difficult for farmers to repay the loans. The FERA screened applicants in an effort to offer loans to those families that would be most likely to repay the funds, making it difficult for the most destitute of families to get federal help. Additionally, this type of relief did not dismantle the sharecropper system, thus perpetuating the problem of southern farmers.[4] The Agricultural Adjustment Administration (AAA) was created in 1933 to address the lingering southern agricultural disaster that beset the region in the mid-1920s.[5] The AAA's grand plan was to "regulate the production, flow, and sale of food and fiber to make supply meet demand and no more. Surpluses would be eliminated. Prices would be determined by what was fair for all. Exploitation and unfair competition would be ended. Antitrust laws would no longer be needed because the government would become the trust."[6] In short, it would be a new deal for farmers.

Initially, FDR's New Deal appeared to accomplish its lofty goals. "Outsiders" who generally shunned, and were shunned by, the South traveled throughout the region sharing information and knowledge. Southerners learned about better farming practices, nutrition, and health care. Government intervention in the southern agricultural economy seemed to stabilize markets. Preparations for TVA dams put thousands of men (and a few women) to work and injected money into the economy. As one historian put it, "by the mid-1930s, millions of southerners had tasted the New Deal tonic, and many had become addicted."[7] It seemed as if the influx of people, ideas, and especially federal dollars "could not fail to have a profound impact on the South. The federal government intervened directly in the daily lives of almost every southerner."[8]

Yet the outsiders assigned to fix the problems were unprepared for both the depths of poverty they witnessed and the difficulties they faced in ameliorating associated problems. "New Deal relief agencies first uncovered the full, hitherto unsuspected, dimensions of rural poverty. Social scientists [at the University of North Carolina] and southerners in federal agencies in Washington pulled this data together to proclaim the South in 1938 as the nation's number-one [sic] economic problem."[9]

Solutions that seemed viable in theory proved nearly unworkable in reality. Landowners resisted changes to the sharecropping system. Farmers balked at plans to leave fields uncultivated. The moneyed class contested perceived threats to the social hierarchy. Powerful corporations challenged government interference in business practices. Racial segregation was codified, entrenched, and intransigent.

As the 1930s progressed, it became increasingly clear that the South's problems were neither easily nor quickly resolved by New Deal programs. They were simply too vast, and proposed solutions challenged too many social, economic, and political foundations in the region.[10] Historians have since revealed fundamental barriers that prohibited profound change during the New Deal era of the 1930s. Some scholars argue that "long-standing attitudes regarding self-reliance [and] limited government" prevented the New Deal from achieving full success.[11] Others assert

that the failure to accomplish significant changes in the sharecropping system challenged the program's overall success.[12] Still others believe that New
Dealers were overwhelmed by the deeply entrenched and profound poverty,
which they inadequately understood and appreciated. Indeed, "much of the
region's poverty was overlooked precisely because it was a long-standing
situation. . . . Thus in the early New Deal only the most perceptive government observers . . . understood that widespread destitution was a problem
distinct from the depression."[13] As one historian succinctly asserted, "the
impact of New Deal policies in the rural South proved complex, contra-
dictory, and revolutionary."[14] In reality, the New Deal proved to be the first
step on the long road toward southern rehabilitation and recovery. World
War II, postwar judicial interventions, and the Civil Rights movement built
on New Deal accomplishments to foster lasting change in the South, and
particularly the considerable economic growth that started in the 1970s.[15]

Of course, in 1933 nobody had the ability to predict that the New Deal
would achieve limited success. Looking instead across a vast swath of the
country and seeing agricultural and economic devastation, New Dealers
swung into action. In north Alabama in particular, the type of assistance
residents needed was far more than the government relief so many of
them already received. Despite their poverty, rural southerners along the
Tennessee River were proud and generally hardworking people, chasing a
dream that seemed unattainable—an escape from the sharecropper system
that shackled them to worn-out land, a life without government assistance,
and the ability to live healthy lives. Loans, while appreciated, were not the
solution. More education, less isolation, and industrial job opportunities
were better solutions than continued dependency on the relief agencies,
which were failing to meet the needs of large farm families with too many
mouths to feed and, ironically, not enough food to feed them. Years before
he became president, FDR traveled through the South and publicly noted
the many social problems facing it, as well as the fact that no one had yet
offered any substantive options to help the South catch up with the rest of
the country.

Historians have written of the compassion FDR felt toward the South,
and his understanding that they did not bring their situation on themselves
but that it was a result of unfortunate circumstances, including failures of
Reconstruction and larger social problems previous presidents failed to
address. A large-scale regional, geographical, and social development program unlike any other ever created, arguably larger than any one private
industry could have ever undertaken, was needed. Along with other New
Deal agencies, the TVA was created to help.

The TVA was different from other New Deal relief projects due to its comprehensive focus on geographic and socioeconomic development. Other programs such as the FSA and FERA offered small-scale, temporary relief to individuals and families. The TVA, on the other hand, promised financial relief for the entire Tennessee River Valley region. The program hinged on the creation of dams along the Tennessee River that would produce cheap electricity for rural residents. Through the powers of eminent domain, the TVA obtained hundreds of thousands of acres of land to create reservoir and recreation areas. The TVA also coordinated with a host of other organizations and agencies, such as Home Extension agents, to personally train farmers on how to best use electricity and new farming techniques that made better, more productive use of the remaining farmland. As the ultimate goal of the TVA was to eliminate poverty, recognized as the root of the socioeconomic problems in the South, public opinion surrounding its efforts was immediately favorable; finally, the government's intervention seemed large enough in scope to make sweeping changes needed for a better way of life.

Despite having official headquarters in Knoxville, Tennessee, the TVA arose primarily out of a failed dam construction project in a small town in northwest Alabama. Muscle Shoals was the site of not only the most dramatic problem with Tennessee River navigation but also the locus of the worst poverty along the river. The origin story of the TVA is one as winding as the Tennessee River itself, involving wars, foreign dependency on nitrates, Henry Ford, senators burned in effigy, death threats, and, ultimately, a dominant government force that remains powerful in the Tennessee Valley today.

Origins of TVA

For generations prior to the Great Depression, Alabama's political leaders made empty promises to deal with the Muscle Shoals "problem."[16] In 1918, construction began on Wilson Dam and a nearby nitrate plant, designed to eliminate the United States's dependency on Chile to supply nitrates needed for war.[17] But the conclusion of World War I essentially halted the dam's construction, and it stood unfinished for over a decade while Congress debated what, exactly, to do with it. There were two viable options: finish construction and put the dam to work or tear it down and sell the scrap materials to recap some of the investment. Congress favored completing construction, recognizing the need for nitrate production for war efforts and fertilizer for the many farmers of the region, and because Wilson Dam held great potential to generate large amounts of hydroelectric power to serve the area's rural community. Progressive Republicans saw the dam as an opportunity to jump-start plans for much-needed regional development. Because farms in north Alabama mostly lacked access to electricity, the dual purposes of fertilizer and hydroelectricity production seemed to some in Congress like a solid plan. The issue would be convincing the executive branch to support such a plan. Private industry might also take over the dam, finish its construction, and make it productive in any number of ways. So, in 1921, Secretary of War John Weeks called for interested parties to make a proposal to Congress about what to do with the unfinished Wilson Dam.

Weeks may have thought that the Alabama Power Company would jump at the chance to procure a structure that would generate so much hydroelectric power, particularly rural power. The private power company did, in fact, have its eye on the Wilson Dam for some time.[18] Alabama Power's primary customer base did not include rural farmers, but they were slowly working to build rural power lines.[19] Ultimately, however, because north Alabama farmers could not afford electricity, the power company did not make a formal offer for the site.[20]

Instead, an unlikely bid came from auto magnate Henry Ford, who petitioned Congress to lease the plant to him for the next 99 years. Ford's proposal generated a substantial amount of public support in north Alabama; he was the first industry giant who seemed interested in developing the same sort of industrial infrastructure in Alabama as was present in the North, Midwest, and New England. Ford promised that, if Congress agreed to the vague terms of his deal, he would create a city "seventy-five miles long" that would serve as one of the country's premier industrial centers.[21] Talk of Ford's plan sparked much enthusiasm, from Lauderdale County to New York City. Newspapers and public opinion praised Ford's deal, which resulted in a premature real estate boom unlike any the area had ever seen. Even though Congress had not yet approved the deal, local residents started preparing for the assured influx of industry promised by Ford's plan. New York real estate investors almost immediately began purchasing massive amounts of cheap land to resell to people across the country at much higher costs, promising these new landowners that their investments—many of them committing entire life savings—would be returned many times over.

Ford sought and received validation and support for his proposal from Thomas Edison, whom he brought along to a dam inspection at Muscle Shoals. Their trip attracted much local attention. To no one's surprise, upon "inspection" of the dam, Edison told Congress that Ford's deal was feasible. This partnership with one of the country's greatest minds added further enthusiasm to Ford's plan and gave the people of north Alabama great hope.

Not everyone supported Ford's proposal, however. His plan was vague and drew criticism from congressmen and noted conservationist Gifford

Pinchot, who argued the plan failed to account for the development of the natural resources of the Tennessee River. Nebraska senator George Norris, a leader of the Republican progressives, led the congressional fight against Ford's plan for more political reasons. As someone who'd tried for years to convince Congress to use the Wilson Dam site as a catalyst for social and geographical change in the South, he saw Ford's plan as a terrible idea. Norris saw the need for more unified government intervention to facilitate the comprehensive development of the Tennessee River region to truly eliminate the poverty and sharecropper system that consistently abused the land. After a lengthy congressional battle, Ford withdrew his offer to lease Wilson Dam, and Norris convinced Congress that a more comprehensive plan was needed. Even though Ford was ultimately responsible for backing out of the deal, many in the South and across the country who'd invested their life savings on land that lost value almost as quickly as it had gained it, unfairly blamed Senator Norris. He received death threats and was burned in effigy. Hatred was so intense that on a visit to Muscle Shoals shortly after Ford withdrew his offer, Senator Norris was assigned an armed guard for his own protection.[22]

Norris eventually found support for his ideas about large-scale regional planning after the Army Corps of Engineers finished a comprehensive study of the Tennessee River region years later in 1931. In their final report, the Corps advocated for the creation of a government program to facilitate the "unified development" of the Tennessee Valley. The report further suggested that development take place gradually as a joint effort between public and private entities. The timing of the report could not have been better. Lengthy congressional fights over Wilson Dam and the ultimate development of the TVA were finally settled with the election of FDR in 1932 who, during his historic first one hundred days in office, passed legislation creating the TVA, much to the delight of Norris and other progressives. Roosevelt referred directly to Wilson Dam in his address to Congress supporting legislation to create the TVA, calling it "a small part of the potential public usefulness of the entire Tennessee River."[23] The TVA would embark on hydroelectric energy, flood control, land improvements, and support of more industry. FDR saw the need for a comprehensive program of development similar to the one urged in the 1931 Corps of Engineers study, with one big difference: development would be exclusively controlled by the public sector via the TVA.

The TVA would be a unique government program, unlike any previously created. FDR called it "a corporation clothed with the power of government but possessed of the flexibility and initiative of private enterprise."[24] One of the most unique characteristics of this agency was its ability to create and approve its own projects.[25] This cut out considerable bureaucratic red tape facing other government agencies and facilitated incredibly fast-moving action, resulting in the TVA starting and completing 20 dams along the Tennessee River in 20 years. It also meant that the TVA had a great deal of inherent power, more than any private corporation. Because of its unique position as a government agency, it had the power of eminent domain, a crucial need when manipulating a landscape.

The TVA Act was signed into law on May 18, 1933. The very next day, a three-man board of directors began to take shape, and employees were quickly hired for high-level positions. FDR hoped the TVA would improve the poverty-stricken southerners' way of life. He cared about the South; many trips to Warm Springs, Georgia, for polio treatments had opened his eyes to the realities of life there.[26] Comprehensive changes to north Alabama residents' way of life would require a comprehensive, multifaceted program addressing many social and economic problems facing rural southerners, one that would "provide for the general welfare of citizens of [the Tennessee Valley] . . . all for the general purpose of fostering an orderly and proper physical, economic, and social development of said areas."[27] The TVA was given congressional authority to create structures and programs that would provide flood control, reforestation, proper use of marginal lands, agricultural and industrial development, and national defense and improve health and recreation.

FDR wanted the TVA to serve as an example to the rest of the world in "tying in industry and agriculture and forestry and flood prevention, tying them into a unified whole over a distance of a thousand miles."[28] Success of these goals hinged on the construction of a series of high dams along the Tennessee River and smaller dams along its tributaries that would, among other things, generate affordable hydroelectric power. That power would then be sold to residents, mostly rural, at a fraction of the cost private companies were charging at the time. The creation of dams would also facilitate the additional TVA goals of improved navigation, flood control, malaria prevention, and improved land use. The issue of power production and distribution generated much controversy, as it marked the first time the federal government had entered the utilities business. Power production was argued later to be a secondary, albeit important, aim of the construction of dams. The construction was justified for two practical purposes: improvement of navigation by ensuring at least a 9-foot channel along the entire river, and the prevention of devastating floods that routinely destroyed farmland. The total estimated cost of the dams in north Alabama was $106 million—roughly $32 million for Wheeler, $29 million for Guntersville, and $45 million for Pickwick.[29] In the eyes of Congress, this was money well spent.

One element of the TVA that made it special was that, by design, it was ready to partner with other relief and government agencies, all working together to rid the South of poverty. This proved to be a team effort that involved a host of other existing and new government agencies that all immediately went to work to carry out the provisions of the TVA Act. Noting the extent to which government agencies worked together to make the TVA's goals reality, early TVA administrator Marguerite Owen recalled the extensive cooperation among a number of governmental agencies, including the Civil Service Commission, the Bureau of Chemistry and Soils, and extension agencies of land grant colleges.[30] In all, twenty-eight federal agencies worked together to support the initial development of TVA programs.[31] Conveniently, the TVA was able to draw on the expertise and resources of several other government entities to jump-start their massive construction projects.

Passage of the TVA Act was also a major political move for FDR, as it helped him solidify support from both southern Democrats and Republican progressives.[32] Though Congress had just become Democratic controlled, Roosevelt was well aware he needed the Democrats' backing as well as that of Republican progressives who had pushed ideas similar to the TVA regarding government-based conservation and development in order to maintain his growing political power. Similar to other New Deal programs, as one biographer suggests, the TVA was a program that FDR created quickly and then delegated to others to manage.[33] Roosevelt's vision for large-scale geographic and regional planning was to be administered by a board of directors, also idealists in their own ways, who saw the need for giving the South a chance to pull itself out of poverty.

Board of Directors

Roosevelt's first appointment to the TVA was the chairman of the board of directors, Arthur E. Morgan, a former president of Antioch College. As an engineer with experience on seventy-five different water control projects, mostly in rural Ohio, Morgan was initially in charge of all construction projects as well as the coordination of land surveys.[34] Immediately upon his appointment to the board of directors, he began surveying and planning for the construction of Norris Dam, the first dam completed by the TVA, then Wheeler, Pickwick, and Guntersville, all of which would have reservoir areas in north Alabama. Morgan apparently was motivated to join the TVA board of directors because his personal vision for the TVA aligned with Roosevelt's. He also saw the need for large-scale, comprehensive planning that centered on social welfare.[35] In writing about the origins of the TVA, he said, "my primary hope for the TVA . . . was that it might create a new spirit and attitude in a public service."[36] An idealist, a visionary, and a big-picture thinker, Morgan had grand plans for the TVA that would soon conflict with those of the other two board appointees.

Harcourt Morgan (no relation to A. E. Morgan) was the second appointee. As the former president of the University of Tennessee, he was the only board member with farm connections and the only one with any firsthand knowledge of the Tennessee Valley.[37] Though he was born in Canada, most viewed him as a southerner because of his long relationship with Knoxville. This was an important characteristic that served him well in dealing with the people of the Tennessee Valley, who trusted him as one of their own. His primary area of responsibility, and primary goal in working with the TVA, was in meeting farmers' needs. In particular, he hoped the TVA would be able to revolutionize fertilizer production for local farmers.[38] He viewed the production and use of fertilizer as one way to improve the plight of the rural southern farmers. A true educator, he felt farmers in the South could improve their practices with education. His passion for the TVA came largely from his "common mooring" idea, a belief that humans and nature were inextricably unified. He was described as having "a concept of nature which was as great as that of Francis Bacon . . . in understanding the relation between land, water, and people."[39] A better orator than writer, Harcourt Morgan's strength was in talking with Tennessee Valley residents about what the TVA would mean for farms, immediately and for generations to come.

The final member to join the board of directors was David E. Lilienthal. A lawyer by trade who had worked extensively with the Wisconsin Public Utilities Commission, he was an excellent choice to be the person responsible for overseeing the hydroelectric power aspect of the TVA, including rural electrification. Despite his experience and a number of well-known supporters—including US Supreme Court Justice Louis Brandeis and Wisconsin governor Philip LaFollette, who had appointed him to the Wisconsin Utility Commission following a landmark Supreme Court victory for public utilities—Lilienthal initially felt he was largely unqualified for the position. At the first official meeting of the board at Chicago's Palmer House hotel, he admitted these feelings to A. E. Morgan. At 32, he was roughly half the age of the other two board members. He was not a Southerner and had never lived or worked in the South; nor was he a farmer or an engineer. But his concerns soon disappeared and transformed into enthusiastic acceptance of the position. His youthful energy and ambition translated into an exceptional work ethic that made him a tireless promoter of the TVA. He found a true mentor, if not "kind of a father," in Harcourt Morgan, who tended to share his philosophy about the TVA more than A. E. Morgan's. Harcourt Morgan's vocal support of Lilienthal, combined with Lilienthal's desire to meet and talk with rural southerners, made him a welcomed figure in many cities across the TVA service area. He once said, "I just like people and I liked the people I met, and if you like people and show that you do, they are likely to be sympathetic with you."[40]

Initially, the board carved out governance territories based on their expertise and experiences and worked individually in each area to get the TVA off the ground. They were productive despite considerable internal strife, a subject that has been extensively researched and documented. Historians have been particularly interested in the tension that arose at that very first meeting at the Palmer House, which ultimately led to a full-length congressional investigation and the forced removal of A. E. Morgan from the board of directors in 1938. The conflict stemmed from fundamentally different perspectives about the agency and the most effective way to implement their far-reaching programs. One key difference in opinion between A. E. Morgan and Lilienthal, for example, was A. E.'s desire to work with, instead of compete against, private power companies. He was convinced they would eventually agree that government power was a better option than private power.[41] A *New York Times* article called A. E. Morgan a "baffling personality," and largely supported his removal from the board.[42]

A former TVA librarian, Mary Utopia Rothrock, neatly summarized the polarization of attitudes forming the crux of the tensions. She described A. E.'s attitude as one of intending to "go down there and uplift" the people of the Tennessee Valley, while Harcourt and Lilienthal's approach was "look, these people are all right and if we just straighten out some of the inequities and just lend a little help they will take care of themselves."[43] A. E.'s top-down approach directly conflicted with the "grassroots" approach preferred by Harcourt and Lilienthal, which ultimately won out as a key descriptor of the TVA. The grassroots effort likely began with Tennessee Valley residents participating in the TVA workforce.

Other TVA Employees and Jobs for the Valley

The Great Depression had one positive impact on the TVA's early hiring procedures: They essentially had their pick of highly educated and talented employees because many of the country's best engineers, researchers, and skilled laborers were out of work or underemployed. People came from all over the country to work for the TVA, resulting in an influx of highly skilled workers to the South. Lilienthal once called the TVA a "dream world of all engineers."[44] It was a "once in a lifetime opportunity" to work for an agency like the TVA.[45] The board of directors drew heavily from their professional networks to fill important positions, particularly Lilienthal, who had an extensive network of qualified lawyers experienced in dealing with public utilities. They were, however, careful to avoid traditional and typical "political appointments." This strategy was unique for a government agency. Decision makers at the TVA did not want just anyone who was recommended; nor were they interested in hiring someone because of a relationship with a political figure. Above all else, the TVA looked for two important qualities in employees. They wanted the most talented workers in their respective fields, and they wanted people with enthusiasm who would become devoted TVA loyalists and work tirelessly to implement this grand experiment in the Tennessee Valley.

In fact, the TVA did not just have employees. They had "crusaders" who all felt that they were engaged in "starting something that was important and was going to do a lot of good for a lot of people."[46] World-renowned TVA photographer Charles Krutch called his coworkers "just the best people in the country."[47] Many initial hires remained with the TVA for 20 to 30 years.[48] Not only did people love their jobs with the TVA, their jobs in many cases became their lives. Some became "intoxicated with . . . employment. . . . We talked TVA weekdays, Saturdays, Sundays, and holidays. It was our entire life. It was not only our vocation, but it was our recreation. We were completely absorbed by it. It was a very exhilarating experience."[49] This sort of loyal devotion, bordering on obsession, was arguably one of the driving forces that made the TVA so successful in such a short amount of time. That level of enthusiasm no doubt contributed to the speed and efficiency with which the TVA completed major construction projects.

The TVA implemented progressive strategies in an attempt to be inclusive during a time when workforce diversity was not a typical agency value. The TVA routinely hired women and black workers. Women worked in traditional roles as stenographers and office assistants, working hard to document the many varied aspects of the TVA project and serving as secretaries for TVA leaders. But others were hired for more visible, even nontraditional roles. For example, Martha Branscombe was hired as a family caseworker to deal with some of the most difficult relocations in north Alabama. TVA leaders originally intended for all caseworkers to be male, but Branscombe's training in social work and her ability to relate to rural southerners set her apart from the team of male coworkers with whom she worked closely. Another example of a woman hired for an important TVA role is Marguerite Owen, who was hired for a highly visible, extremely powerful administrative position in the TVA's Washington, DC, office. Hand-selected by Lilienthal, he felt that "a woman could handle a Congressman and say 'no' a lot better than a man. It is sometimes more difficult for a predatory patronage hunter to deal with a woman than a man."[50] Owen spent a lengthy career with the TVA, as did many other women who were hired in the earliest days of the agency's existence.

While the TVA hired black workers on a quota system meant to compose a workforce of the same racial proportion as the community in which they worked, multiple factors prevented the TVA from helping to desegregate parts of the rural South. One account suggests that racist attitudes among local communities hindered the TVA's ability to treat black workers fairly. Blacks were "discouraged and prevented from taking the examination to work at Wheeler Dam, not because of any official policy of TVA, but because . . . local postal employees . . . refused to give applications to blacks. The TVA elected to waive the examination requirement for blacks rather than supersede or disrupt local control over the distribution process."[51] Yet the TVA still largely participated in segregationist policies. The communities that formed around the dam construction sites were heavily segregated, with separate living facilities for black and white workers. Restrooms and water fountains were segregated, remaining that way for many years. It was not until former Prime Minister of India Jawaharlal Nehru planned to visit Norris Dam that a TVA photographer insisted signs declaring the segregation of facilities be removed. It is interesting to note that the photographer had difficulty explaining to fellow employees why it would reflect poorly upon the agency.[52]

Their employment quota system and segregated living quarters did not improve race relations in the Deep South, especially in light of the tense and ongoing racially charged trials of nine black men in nearby Scottsboro. Instead, their treatment of blacks served as what the TVA saw as a fair compromise—a way to escape harsh public criticism from outsiders who may have been quick to blame the TVA for participating in a segregationist system. This was as much a political move on the TVA's part as it was anything else. FDR was notorious for ignoring race relations in the South in order to keep the political support of powerful, segregationist white southerners.[53] A later investigation conducted by the National Association for the Advancement of Colored People of the TVA's policies toward race relations resulted in a critical report outlining the "lily white" policies of the TVA, despite the agency's claims that their willingness to hire black workers proved they were progressive in the area of race relations.

The TVA's employment practices, then, can be both praised and criticized for the manner in which they treated women and minorities. This aspect of TVA history is as complicated as the process of bringing rural electricity to thousands of homes across the South. Women were hired for some important, even high-profile positions, but most women who worked for the TVA held clerical and administrative positions, consistent with traditional job duties for women in the workplace at that time. Most jobs, including immediate positions available in the areas of land clearing and manual labor, were intended for men; women were rarely, if ever, hired for those positions. Black people had some opportunities to work for the TVA, but life in the Jim Crow South made those opportunities more difficult to take advantage of.

Rural white men were heavily recruited by the TVA for skilled labor positions. Consistent with their grassroots approach of putting the people of the Tennessee Valley back to work for themselves, local residents were among the very first hired by the TVA to do the grunt work needed to make way for a series of dams. Though north Alabama residents were excited about the possibility of Henry Ford's investment in their future, they were truly overjoyed with the news of FDR's more comprehensive plans. The promise of new jobs was one of the reasons north Alabamians so readily embraced the new government program. FDR had, after all, said the initial purpose of the TVA was to "put Muscle Shoals to work."[54] This was music to the ears of many Tennessee Valley residents. Both skilled and unskilled workers would be needed to make the TVA function, and a considerable number of those workers would potentially come from the communities in which TVA would soon serve, an approach that fit the greater mission of FDR's New Deal programs.[55] Putting residents to work in nonagricultural occupations increased the chances of eliminating the sharecropper system that perpetuated poverty. More jobs meant more money in the pockets of cash-strapped families, and FDR promised that comprehensive regional development would surely result in increased industry, meaning more non-agricultural jobs were just a few years away. Those who wanted to stay in farming would likely earn bigger profits than before, with elimination of the threat of erosion from flooding and increased education about best farming practices.

The work was hard, and many of the jobs were temporary, such as land clearing efforts, but that did not deter people from embracing opportunities for good-paying jobs. The TVA brought so many jobs to the area that "at times a third of the active working force of the Tennessee Valley has been engaged in clearing timber from reservoir sites."[56] Jobs such as truck drivers, tree climbers, boat operators, machine operators, tractor operators, and carpenters were well suited for farmers who so desperately desired alternate ways to make a living. The pay was good, too. Day laborers earned 45¢ per hour, while dragline operators could earn up to $1 per hour.[57]

Rural Electricity, Yardstick Rates, and the Power Struggle

The TVA's plan to bring electricity to rural areas hinged partly on local communities along the Tennessee River voting to give their municipalities permission to receive and distribute TVA power. In other words, for the TVA to work, north Alabamians and others had to welcome TVA electricity into their homes and become enthusiastic users of the new power source. If the TVA failed in north Alabama—the site of three large dams and one additional reservoir area—the entire TVA project would be in jeopardy, and the government would lose millions of dollars already spent on construction efforts. But if north Alabama readily adopted TVA power, it would highlight the agency as an example of what could be achieved with cheap electricity and large-scale regional planning and development. With this in mind, Roosevelt alluded to his eventual plan to develop other TVA-type agencies across the country. He once commented, "If we are successful here, we can march on, step by step, in a like development of other great natural territorial units within our borders."[58]

This aspect of the TVA project was brought into the larger "grassroots" narrative championed by board member David Lilienthal.[59] His belief about how the TVA would be most successful mirrored the attitude of many rural southerners, who wanted to help themselves out of poverty rather than to be lifted up by a large, powerful government entity. He wrote extensively of the many aspects of the TVA program that were inherently grassroots oriented, from adopting electricity to farmers educating their neighbors about better fertilizer practices.[60] However, this approach opened the TVA up to criticism that it was socialist in nature. There are some clear parallels between the TVA and socialist ideals, such as the community housing the TVA built for workers and decisions about land use and regional planning made at the federal level for the supposed good of all the region's residents. Congressional critics argued that the TVA was "patterned closely after one of the soviet [sic] dreams."[61] Even the *New York Times*, which later supported the TVA, initially asserted that "enactment of any such bill at this time would mark the 'low' of Congressional folly."[62] Board Chairman A. E. Morgan defended the TVA as a completely American ideal, saying, "It won't be capitalism; it won't be socialism; it won't be individualism; it won't be any of those *isms* that have become little more than labels or battle slogans. . . . It will be a new Americanism."[63]

Early TVA employees, particularly a handful working in Knoxville head-quarters, were found to have direct ties to the Communist party, which later led to multiple investigations by the Federal Bureau of Investigation.[64] However, for the most part, TVA employees denied the connection. Dr. Earle Draper recalled how he would counter the critique many made that the TVA was socialist, particularly as it pertained to land acquisition. He stood by his belief that the TVA was not socialist when "the main objective was to improve the economic status and to transfer the status of people from dependence on such welfare as there was to a situation where they made a decent living."[65] The parallels between the TVA and socialist ideals, especially at that time in history, are undeniable; however, many historians have agreed that despite their criticisms, New Deal plans were not socialist in nature.[66]

Any socialist notions were readily overlooked by rural Alabama residents, who were desperate for electricity that Alabama Power was unable to provide, electricity that they could not have afforded even if it was available to them.[67] The irony of living in the shadow of Wilson Dam yet having no electricity was not lost on Fannon Beauchamp, a division engineer at the dam: "Within 3 miles of Wilson Dam [one] could hear the roar all night, and that little community never got power until TVA built out to them 3 miles [away]."[68] People wanted electricity but had no way to get it and no money with which to purchase it. This led to another reason many communities readily embraced the TVA's presence. The TVA established "yard-stick" rates for offering electricity at prices drastically lower than private power companies like Alabama Power. The TVA described their rates as "yardstick" because they intended for their lower rates to be those by which private power rates were measured.

Together with a group of trusted experts, Lilienthal took the lead on re-searching and planning the per-kilowatt hour charge to TVA customers. The rates hinged on the amount of electricity consumed. The more residents used, the cheaper their rates became. TVA rates were a fraction of rates offered by private power companies like Alabama Power. This provided a dual incentive to rural residents to consider voting to allow TVA power in their municipalities: rates were cheap, and there was motivation to use more electricity. Though the TVA wanted its electricity to be afford-able, they no doubt expected that lowering rates in this area would result in lower rates across the country and across other areas of the South not served by TVA power, hence the "yardstick" terminology. Private power companies took notice and were understandably unhappy. In a thorough history of the Alabama Power Company, Atkins says that Alabama Power was the first, not the TVA, to suggest the concept of yardstick rates, but at that point in time, no other area of the country fully adopted them. The obvious drawback for private power companies using cheap rates was the negative impact on profits.

One obvious problem remained, even with affordable electricity rates. How could rural residents, many of them farmers barely scraping by, afford the appliances needed to take advantage of the newly available electricity? Yet again, David Lilienthal had a creative idea for jump-starting the economy and encouraging rural residents to use electricity. His prior experience in Wisconsin positioned him well to negotiate with major manufacturers General Electric and Westinghouse to build affordable appliances of the same quality and dependability as their more expensive models. Lilienthal also spearheaded the creation of the Electric Home and Farm Authority (EHFA), through which the purchase of electric appliances was subsidized and financed for low-income families across the South. The EHFA resulted in a considerable jump in ownership of home electric appliances in 1934, and Tennessee and Georgia were the top two states purchasing ranges and refrigerators in the entire country that year, which helped to dramatically increase the amount of electricity consumed by residents.[69] Placement of electric appliance showrooms in local stores allowed the TVA to orchestrate events such as cooking schools to educate residents, mostly housewives, on how to use the new appliances and contributed to the jump in sales.[70] The EHFA also funded newspaper advertisements of many electric appliances around the time of the TVA's presence in north Alabama, further adding to the discourse about the changes the agency would make in the Tennessee Valley.

Unfortunately, the success of the EHFA was short lived. It turned out to be an insignificant force in sales, likely due to the fact that in the earliest TVA years, the poverty-stricken residents of the rural South were still unable to afford to buy appliances.[71] However, the EHFA did play an important role in convincing some to use electricity. Those who could afford it tended to take advantage of the program, which initially helped the TVA sell more electricity. It was a solid attempt at economic expansion, all part of the New Deal's plan for economic improvement.[72] Though not as long-lasting or successful as Lilienthal's other initiatives within the TVA, the EHFA is an example of his ability to work with a professional network of industry leaders, formed during his time in Wisconsin, and the creativity with which TVA employees often approached their roles.

North Alabama residents famously and overwhelmingly voted to start using TVA power, which meant a discontinuation of Alabama Power for their electricity needs. Understandably, the TVA's presence and positive response among consumers did not sit well with Alabama Power or its board of directors. It led to a lengthy legal battle between public and private utility ownership that ultimately questioned the constitutionality of the government entering the power business. It is interesting to note that earlier legislation, similar to the act that would ultimately create the TVA, also evoked this controversy. Norris's earlier plans for generating and selling power at Wilson Dam failed under the two previous probusiness presidents, Calvin Coolidge and Herbert Hoover, partly because of their resistance to enter the government into business ventures. Hoover said that the government producing electricity "would be the negation of the ideals upon which our civilization has been based."[73] FDR, however, firmly believed that government's provision of low-cost electricity was a cornerstone of comprehensive regional development in the rural South.

The TVA could maintain low power rates in part because, as a government program, it could avoid paying some of the major taxes public corporations were required to pay. Under Lilienthal's direction, the TVA purchased Alabama Power's distribution lines and other necessary equipment in north Alabama, which essentially nudged Alabama Power out of the north Alabama market. Alabama Power's trustees eventually filed suit against the move the distribution line and equipment sale, but they also attempted a grassroots strategy of their own to prevent the TVA from supplying power to the rural poor. "Spite lines" often arose in communities that had recently formed electric cooperatives to encourage TVA power. As soon as it was made public that a community had reached an agreement for TVA power lines in their area, according to one TVA engineer, private power companies "immediately threw in construction crews that were to build [spite] lines . . . to discourage any construction on the part of the co-ops."[74]

The legal battles over the TVA's ability to provide power occurred early in the TVA years; "during first five years of its existence, TVA's constitutional authority was questioned in fifty-seven cases and operations were impeded by twenty-six injunctions, of which all but one were later dissolved."[75] In 1936 alone, nineteen power companies filed suit against the TVA claiming damages of $300 million. During these legal battles, particularly the ones dealing with the TVA's right to produce and sell electricity, virtually every other aspect of the TVA moved forward.

Newspapers frequently reported updates on the Alabama Power Company's legal battle with the TVA. Mostly confined to anti-TVA advertisements in north Alabama, a series of critical ads run by Alabama Power attempted to generate public support for privately held utilities, especially surrounding the very public court battles challenging the TVA's constitutionality. Alabama Power used Reddy Kilowatt, a character it had designed to get people to use electricity and to reduce fears associated with electric current, in a 1939 advertisement offering $5 to the top one hundred letters submitted about "what consumers get beyond their power bill."[76] Their closing slogan on the advertisement was "Alabama Power Company: A Private Agency for the Public Good."[77] The *Albertville Herald* ran a series of ads from Alabama Power in 1936, around the time construction on the nearby Guntersville Dam was starting. An advertisement asked residents to strongly consider if they were willing to enter into a 20-year contract with the TVA.[78] They also accused the TVA of "raiding" and disabling Alabama Power's rural electrification efforts. The same ad accused the TVA of discrediting Alabama Power and forcing people into submission: "Federal, State and County employees were made to believe that they would lose their jobs if they signed up with the Power Company. . . . One school teacher was fired for not opposing the Company's activities in the community. . . . TVA will deny this." Alabama Power also accused the TVA of "paint[ing] a glowing picture of the advantages to farmers of organizing into a TVA dominated association engaging in the electric light and power business. Why do they not state the disadvantages?"[79]

In 1935, Birmingham Judge William Grubb caused a scare among north Alabama residents when he ruled in favor of Alabama Power in *Ashwander v. TVA*, claiming that the government did not have the right to produce and sell electricity. The ruling, which threatened the entire TVA project, was quickly overturned by the US Supreme Court, whose members ruled that electricity was a by-product of the dams, not the sole reason for the TVA's existence.[80] The *Albertville Herald* summed up the reaction of many in the Tennessee Valley to the news of the Supreme Court ruling, saying "Valley towns go wild. . . . Joyous greetings echoed and re-echoed up and down the river from Pickwick Landing to Knoxville as residents of this New Deal empire gave thanks that the US Supreme Court had blessed the Tennessee Valley Authority."[81] The role of local newspapers in this aspect of the TVA was crucial; editorials and news stories were written to make the TVA seem like a better option than Alabama Power. And it was hard to argue with that when the TVA's rates were considerably lower than those offered by Alabama Power. The newspapers also reminded readers of just how quickly the TVA was working to electrify the countryside. Frequent articles about powerline construction let residents know the TVA was serious about their commitment to bring electricity to rural areas. Headlines in the *New York Times* following the ruling described a joyous scene across the South, as the "Tennessee Valley shouts with joy. In Knoxville, whistles shriek and band plays while crowd cheers TVA ruling. Muscle Shoals is happy."[82] The

Supreme Court ruling paved the way for the TVA to continue its comprehensive regional planning efforts, massive construction projects, and economic reformation.

Conclusion

TVA was unique, progressive, and an experiment. Nothing like it had ever been attempted in US history, and there was no real guarantee that it would work to fulfill the many promises it made the people of the Tennessee Valley. From its early conception toward the end of World War I as a nitrate plant in the northwest corner of north Alabama to the large government agency–corporation it became, by the early 1930s, the TVA was poised to transform life in the South, especially for the rural poor. Its position as a government agency with the ability to act as a private industry afforded it an extraordinary amount of power, arguably greater than that of any other New Deal program or government agency at the time. Its standing as a private entity entitled the board of directors and other employees in higher-level positions the ability to make important decisions without waiting on the approval of Congress. And its position as a government agency gave it one crucial power that private corporations would not have had: the power of eminent domain. Though the TVA Act gave the charge and set the scope for what the agency was expected to do in the South, the directors had the ability, and the responsibility, to figure out exactly how those goals would be achieved. They quickly divided the labor among the three of them and created concrete plans for reshaping the South, primarily centering on the dam construction and the distribution of cheap electricity to rural families. They hired employees who became TVA loyalists. They also employed the labor force in the communities they would soon serve, offering residents the opportunity for new jobs and allowing them to feel more directly connected to the large government entity.

Before the TVA could put Tennessee Valley residents to work, before it could construct dams, and before it could affect real social and regional change, it needed a considerable amount of land—land currently occupied by rural families. Throughout north Alabama, roughly 2,500 families were living on land that would soon be purchased or condemned by the TVA. The TVA had the budget to buy all the necessary land, and TVA engineers had the power to determine exactly what land they needed and what land would be purchased as a buffer between the reservoir areas and the remaining farms. So north Alabamians, who were among the first to leave the land along the river, crucial in the adoption of TVA power, were also the ones standing in the way of the project's success. The way the TVA handled land acquisition and, more importantly, the relocation of families from that land, would largely determine how the TVA's presence was viewed among residents.

Modernity has a price. The loss of thousands of acres of farmland across north Alabama, even though it was arguably of poor quality due to years of mistreatment, meant a land shortage unlike any ever seen in the area. It also meant the mass exodus of families from the existing riverbanks to higher ground. The group of employees responsible for relocating families from soon-to-be-created dam reservoirs had a taxing job, among the most emotionally intense of all TVA work. If they were unable to successfully convince thousands of people to leave their homes ahead of the rising waters, the entire TVA project would be in jeopardy. Interestingly, constructing the near-ten-story-tall dams proved to be the most straightforward part of the process. Getting TVA electricity in the hands of rural southerners who could not previously afford it turned out, in many ways, to be a greater challenge.

3

Family Removal in North Alabama

In 1935, Martha Branscombe was a relatively recent graduate of the Florida State College for Women. A Union Springs, Alabama, native with a degree in modern languages and a few years of experience working as the assistant to the speaker of the Alabama House of Representatives, she had sufficient credentials for the job of TVA case worker, a position that would have her driving and walking out to rural homes along the river, on land TVA purchased, telling residents they had to move. She had a good knowledge of the life and habits of rural north Alabamians thanks to her upbringing in a small south Alabama town with similar demographics. There was only one thing working against her potential hire: she was a woman, and the TVA expected that only men would be suited to visit reservoir families and ensure they relocated off TVA property.[1]

Branscombe was hired specifically to work in north Alabama with what were known as the "special cases" of family relocation, those families who presented unusually difficult circumstances to the removal effort. Families like Isaac and Ruby Clyde Filmore, who, along with their eight children and Isaac's 66-year-old father, were lifelong sharecroppers in Marshall County, Alabama, living in a home that was "unfit for habitation."[2] Inside the home, Branscombe found "the interior was filthly [sic], swarming with flies and was poorly furnished. Everything was in complete disorder. There were no screens. The small space around the building which sits in the middle of a field was cluttered with rubbish, junk and papers." At the time of Branscombe's visit, Ruby Filmore was suffering from malaria, lying in bed and "unable to sit up."[3]

Clearly, the Filmore family needed help, not only with moving from their home but also with improving their health, housing, and financial situation. And the TVA could help only if the Filmores, along with 15,000 other families living along the Tennessee River, would move. While the TVA paid fair market value for the land they needed, families like the Filmores who did not own their home received no money to move, and no financial assistance to relocate. Their best option was to dismantle the structure they called home and rebuild it elsewhere.

Martha Branscombe had her doubts that the home would survive the move, given its state of disrepair. As she suspected, when the Filmores attempted to move the house, the "lean-to completely collapsed and the foundation gave away shortly . . . and the floor had fallen in. It appeared a hopeless wreck." The family resorted to living outside temporarily until they could figure out a better solution. Ruby Filmore was "gravely disturbed over the situation and the inconveniences and discomfort of being out doors. She was lying on a cot under a fruit tree in the yard." A TVA worker and contractor took responsibility for relocating the Filmores' home, and the family "seemed to rejoice that they had salvaged the ruins." When Branscombe made her final visit to the home just 3 days after her first visit, she found Isaac "in much better spirits . . . certain the house could be used until fall." Branscombe recommended the local department of public welfare follow up with the family because the land purchase and relocation had left them without crops for the coming year.

The Filmore family's experience illustrates the complexity of the problems facing families who had to relocate for the TVA. According to one TVA report, 84 percent of all relocated families needed some form of assistance with the process, ranging from advice about moving to the type of "intensive service" Branscombe assisted in providing.[4] Illness and poverty, combined with poor housing, made relocations difficult. Like many other renters or squatters, the best option was to rebuild their homes out of the same materials on different land the TVA had not acquired. The homes they reconstructed would likely not survive more than a few months. Families would have to rebuild yet again or move on to another home, which was made exceptionally difficult due to the lack of land and housing brought about by TVA land purchases.

Perhaps even more emotionally difficult was the day Branscombe had to take five children aged 4–12 away from their father to the Baptist Orphanage in Troy, Alabama. The children's mother had recently left the family—records show she likely left because her husband was abusive—and the court order involved in this case cited the father's "incompetency and neglect" as justification for their movement to the orphanage. The children had lived as squatters in an 8-foot-by-12-foot tent in a wooded area near Decatur. Unable to afford food or medicine, the children routinely went out in the community to beg for food. When the children got sick, they were given saltwater instead of medicine. The family's only income was the

$8-10 relief given to them by the state each month. While the orphanage perhaps was the best chance for the children to have a better life, it undoubtedly must have been a difficult moment and a difficult car ride several hours away from the only home they'd ever known.[5]

Branscombe joined a team of caseworkers who spent 3 years in the area ensuring that every living soul occupying land that would soon be under water had moved elsewhere. Whatever Branscombe may have been told before signing on with the TVA, and despite what she may have seen growing up in the midsouth region of Alabama, it's hard to imagine she would have been prepared for the intensely poverty-stricken conditions in which reservoir families lived—and had lived in for years with no relief in sight. Trained to help, yet also loyal to the TVA cause, did she question if her work was helping or hurting those with whom she came in contact? What was she supposed to do to help the tenants and sharecroppers who received no money for land acquisition, lost the farmland they had worked, and had nowhere to go after the TVA moved them off their property?

The issue of family removal during TVA construction has been romanticized in Hollywood depictions and historical novels. When picturing what family removal must have looked like, the popular 1966 Montgomery Clift movie *Wild River* conjured a fictitious "Granny" sitting on her front porch with a shotgun, daring law enforcement to physically remove her from her home. In reality, though, family removal rarely worked that way. Some families accepted their fate, and their payment from TVA, peacefully leaving to find other places to live. However, because homeownership and landownership levels were so low in north Alabama, and because dependency on the sharecropper system was so high, most residents received no financial compensation for relocating. Understandably, these families experienced more and different difficulties than their land-owning neighbors.

Branscombe and the rest of the team charged with relocating families were in a precarious position dealing with a most delicate issue. On one hand, they along with their black Ford cars with TVA tags on the front represented the potential for a better life and, as such, could be seen as benevolent helpers. Caseworkers like Branscombe had the ability to coordinate with other relief agencies and ensure that families who moved were in situations that were at least similar to the ones they left. On the other hand, they represented a big government agency that had the power to evict, using legal force to remove people. It's understandable why some may have been intimidated, frightened, or suspicious of the TVA's sudden presence in the Tennessee Valley, especially considering the way it carried out the process of acquiring land and removing families.

Process of Removal and the Land Acquisition Division

Family removal was a lengthy process, but in north Alabama it was completed efficiently and effectively. It began with the Land Acquisition Division using land surveys previously conducted by the Army Corps of Engineers. From these surveys, engineers determined how much land and specifically what land was needed for reservoir creation and the buffer zones around the reservoirs, which were to be used for recreational use or preservation efforts. Next, TVA land appraisers determined fair market value for the land they would need to purchase from owners. After the division had this information, landowners and families on the needed tracts were contacted regarding purchase agreements and relocation. Specially trained caseworkers then entered the area to ensure the families who were required to move eventually left. Relocations and dam construction took place simultaneously. Initial land-clearing and construction efforts began even before residents had vacated the land. The TVA did not have a clear-cut policy for when eviction procedures would begin, instead relying on caseworkers to update them on individual families who were in need of special attention and/or assistance in relocating.

Securing enough land for their massive building efforts was among the critical first steps for the TVA. The Land Acquisition Division was ultimately responsible for determining what land to purchase.[6] The division was also responsible for the "case study, interpretation and service" of landowners, tenants, and sharecroppers directly impacted by the TVA.[7] Although A. E. Morgan described this area of land as "marginal" on more than one occasion, it was worked as farmland by valley farmers. In most cases, land adjacent to the river was quite fertile but also the most subject to devastating floods. Drier, higher land just above the river tended to be much poorer quality soil, overworked and underproductive due to generations of poor farming practices and severe erosion.

The TVA acquired two types of land: land essential to the TVA, for dams and other power structures, and land that was not essential but desirable to facilitate creation of public recreation areas, game refuges, and reservoir buffer zones.[8] The amount of land the TVA acquired served as a point of contention among the board of directors as well as the TVA divisions they represented. Harcourt Morgan, representing the agricultural division, disagreed with A. E. and the engineering division's policy to take more land than the TVA needed to facilitate dam construction. He did not want to take land for recreational and environmental purposes. Harcourt "always resisted, sometimes within reason" the taking of land that was not absolutely necessary.[9] A. E. saw in the TVA an opportunity to improve Tennessee Valley land for the public welfare; his belief in the "good" of government guidance of misused land was a major reason he had agreed to serve on the board of directors. He felt the overworked, marginal land should be placed in better-trained, better-educated hands. He once wrote, "farmers who misuse their land should have it taken away from them by state governments and given to someone who would take care of it."[10] This was, of course, dismissive of Tennessee Valley farmers' lack of knowledge of better farming practices and lack of access to quality education. His view demonstrated a lack of understanding of the plight of southern farmers, for whom the issue of land use and management was not simply one requiring better training and education. Instead, farmers who were misusing the land were generally doing so because they felt they had no other choice but to plant cotton, year after year, on the same soil.

One TVA employee explained that the land acquisition division faced the "complete hostility of the agricultural group" led by Harcourt Morgan, who felt the TVA should purchase only land needed to create reservoir areas. The agricultural group felt excessive land purchase hurt small-scale farmers, even though land planner Earle Draper claimed this was untrue. He claimed that the TVA "simply wanted to enhance the economic opportunities available for the small farmer in the area."[11] Any land the TVA did not use specifically for the construction of dams or reservoir areas might be used for recreation or even dedicated for hunting and fishing. Ultimately, A. E. Morgan's view on land acquisition won out over Harcourt's, resulting in the TVA's procurement of 103,500 acres in the Wheeler reservoir,[12] 110,145 acres in the Guntersville reservoir area, and 63,000 acres of land in the Pickwick area, 23,780 acres of which was in north Alabama. In total, 237,425 acres of north Alabama land went to the TVA.[12] While this was technically more land than was needed for the construction of reservoir areas, part of the TVA's grand plans for the Tennessee River valley region was to engage in land preservation and provide recreation. The TVA anticipated building a buffer zone around the reservoirs that would also provide for recreation and land preservation. This land was taken off county tax rolls and put in the hands of the TVA to use as they saw fit. The TVA's acquisition of this land created a shortage of land available to valley farmers and residents, which proved problematic in north Alabama as families started relocating to make way for Wheeler, Guntersville, and Pickwick Dams.

The TVA made purchasing decisions based on extensive geographical and engineering-based research. They consulted with a host of professionals in a variety of fields, including economists, social scientists, and architects, "all in a group making studies for guidance in the development of TVA to enhance the living standards."[14] Once the TVA determined what land it needed, appraisers were sent in to determine fair market value. Though they worked for the TVA, they were instructed to be objective with their appraisals. Rather than paying a blanket price per acre, the value for each tract was based mostly on the quality of the soil. When appraising land, the TVA took into consideration only the value of the land itself, excluding the value of structures and crops yet to be harvested.[15] Land values were assessed on a scale of seven different "quality grades, ranging from alluvial river bottom to steep hillside. Individual appraisals were made based on a weighted sum of the seven land grades and improvements made to the property."[16] The TVA aimed to buy land at fair prices that would "enable the owners to relocate and reestablish themselves in situations which will afford them equal contentment to that which they now enjoy."[17] Land buyers operating on behalf of the TVA were also instructed to work cooperatively with the owners and "not to take advantage of distressed financial conditions" they may have been experiencing.[18] The TVA did not always buy entire farms, but instead purchased only the amount of land they determined they needed, a policy that could work for or against the landowner. If TVA purchased only a small tract of a farmer's land, then theoretically, the farmer's family could continue a way of life similar to that they experienced prior to the TVA construction. However, if the TVA purchased most but not all of a farm, the farmer may have been left with a tract of land too small to act as a productive plot of farmland.[19]

The TVA employees charged with acquiring land were trained on how to handle the delicate situation of dealing with the rural, poor, marginalized population in north Alabama. The TVA *Instructions to Land Buyers* reminded employees that their job was to help Tennessee Valley residents, keeping in mind the purpose and aims of the TVA. Yet they also served as ambassadors of goodwill, ensuring the public support that was crucial to TVA programs. Especially when it came to the issue of land acquisition, the TVA felt that, "public relations are necessarily of paramount importance."[20] The *Instructions to Land Buyers* warned employees that any poor publicity could be detrimental, going so far as to say negative attention could threaten the entire TVA project.[21] This stern warning pressured land buyers to handle each interaction with a landowner in the most positive way possible, lest the interaction create public scorn for the TVA.

Once the price per acre was determined for a tract of land, the land buyers approached owners with the ultimate price the TVA would pay, averaging $44.11 per acre in the Wheeler Dam area, $50.41 per acre in Guntersville, and $57.31 in Pickwick.[22] Land buyers were explicitly prohibited from negotiating land values, so these figures represented the TVA's "final" offer.[23]

At that point, the owner had two options: either accept the initial offer and sell the land to the TVA, or reject the offer and risk the TVA condemning the land so it could essentially take the land without paying the owner.[24]

The only time TVA reconsidered an offer was if the landowner could verify that the appraiser had made an error in his appraisal, which rarely happened—"once in a blue moon," according to one early employee.[25] This policy, called "no-price trading," was unique, as the TVA was the first government agency to employ this innovation.[26] If the owner did not agree to the price, the TVA could legally begin condemnation procedures.[27] Consider this example of the TVA's immense authority. The TVA had the backing and funding of the federal government, the ability to operate independently as a private organization, and the use of their own internal legal and land appraisal departments to determine how much land they would take for the program and how much they would pay. This "positive coercive power"[28] to acquire land must have been an ominous concept to the poor farmer, who may have felt helpless in the face of this New Deal agency promising to deliver the south from its poverty. Landowners must have felt that failure to sell their land meant they were standing in the way of progress and a presumably better way of life for future generations, not to mention understanding they could lose out on a sum of money unlike any they'd seen before. Likely due to the fear of losing a payout for their land, there was little legal resistance among landowners and no forced evictions in either the Wheeler or Guntersville reservoir areas. However, a small percentage (less than 10 percent) of the land purchased in north Alabama was condemned.[29]

The sufferings of relocated families are illustrated by the story of Henry "Buddy" Cole. Cole had a successful painting and carpentry business in nearby Huntsville, Alabama, until an infection caused him to lose both of his legs just below the hip. Unable to continue his trade, but being a "thrifty, hard-working man" who did not want to be dependent on anyone else, he built a house on land borrowed from an owner named Mr. Walling, who lived adjacent to the river in the Wheeler reservoir area. Walling did not give Cole a deed to the house or any land, so when Mr. Walling passed away, there was no written or verbal record of his legal right to stay on the land. For that reason, the land on which he lived was condemned by the TVA. When Margaret Branscombe, the TVA's female caseworker, visited Cole, she found him "very bitter because of the loss of his house," yet he claimed he did not feel antagonistic toward the TVA.[30] His house was set up to provide easy access to the river for him to conduct his commercial fishing operation. As Branscombe described it, "from the front porch there is a slide constructed of lumber which goes to the water's edge. By means of this slide, Cole is able to get into his boat, which is his only means of transportation." Cole sold his fish for a small profit with the help of his 17-year old son, who had epilepsy and needed medication, who would take the fish to nearby Huntsville on his bicycle, selling it for 15¢ per pound.

Unable to convince the TVA to allow him stay in his current location, Cole wrote letters to Alabama Governor Bibb Graves and President Roosevelt. His emotionally appealing letter to FDR said, "I am a poor man on top of that I have had the misfortune to lose both of my legs and have a wife and 2 children. You know a man with both legs off can't do much to make a living. . . . I had to move out of town on the count of not getting no work. . . . I have been living heare [sic] 4 years and the TVA has cut the bank off an [sic] bought the land and sent me word to move by the first of March. Mr. Roosevelt, if you are helping the poor please help me; the relief is helping me some now but not very much but I am thankful to get that. . . . I cant [sic] pay rent and the relief won't pay rent and buy coal."[31] The letters got him nowhere. Branscombe's ongoing visits finally convinced him that moving was inevitable. He eventually resolved to build himself a houseboat, which was the best option he saw for continuing his fishing operation, his only source of income. He and his family moved into the houseboat on November 18, 1936, at which point Branscombe found that his attitude toward the TVA had shifted to "rather critical . . . defiant and resentful."

A total of 842 families in the Wheeler reservoir area, 1,182 families in the Guntersville reservoir area, and 499 families in the Pickwick reservoir area were required to give blind consent and compelled to believe that what the TVA was doing was in their best interest, and in the best interest of future generations. For landowners, there were direct monetary benefits to agreeing to the TVA's land purchase price. Those who sold their land received more than they may have ever received in one lump sum up to that point in their lives. Luther Tidwell, who was 12 years old when the TVA bought his father's 144-acre farm, distinctly remembers that his family received $9,999.99 in exchange for their land, which "was a lot of money for back then," and higher than the average cost per acre for land in Guntersville.[32] The Tidwell family was lucky. It was enough for Luther's father to purchase another farm in the area that remained active for the next six decades and on which part of the family continues to live. Finding land of equal quality and size at the time was difficult for them, though, as it was for so many other families.

The TVA not only acquired farmland. They also took possession of several graveyards. While the land was razed of all plant life, homes, and other structures before flooding took place, determining what to do with the graves proved to be a much more delicate and different issue. Honoring the dead was understandably an important community ritual, and in north

Alabama during the 1930s, many people paid frequent visits to graves to pay respects to deceased friends and family. Still today in this region, cemetery decoration days are an important part of many communities and serve as a time for fellowship, interaction, and remembrance of deceased family and friends. Originally, the TVA had planned to build a national cemetery near Norris Dam, to which remains would be moved.[33] However, this decision did not sit well with the residents of what would become the Norris reservoir area. In response to the community's protests, the TVA changed their policy to contact the next of kin for every identifiable grave, allowing him or her to determine whether to leave the grave in place or have remains removed. This policy was repeated in the Wheeler, Guntersville, and Pickwick reservoir areas. While many graves were relocated at the request of families, some graves remain inaccessible today.

Before the TVA established a presence in north Alabama, neighbors to the north in Tennessee were already experiencing the potentially negative effects of large-scale land acquisition. The construction of one TVA project meant flooding "a small village . . . scores of cemeteries, churches, schools, and farmers' homes . . . tracks of the Southern and the Louisville & Nashville railroads, parts of two main highways, county roads, several bridges" and "the historic home of General Joseph A. Cooper, famous Tennessean."[34] Despite the potential drawbacks, the extent of land and architectural structures lost, and the hardships families faced from forced relocation, the TVA carried out their land purchase and construction projects with relatively little resistance among residents.

Land acquisition and the subsequent removal of families took place in phases for each dam and reservoir area. The most immediate need was for those who lived in the area where the dam would be built to leave, usually 1 or 2 months after the TVA took ownership of the land.[35] Residents living farther out from the reservoir area and dam construction sites could take longer to leave because their land would not be flooded until after the dam was finished. The TVA had the authority to use force to move structures and people if they stood in the way of the agency's plans.[36] However, the actual use of force could prove problematic for the TVA, as the agency hoped to remove people as quietly as possible. Forced removals could have constituted a newsworthy event picked up by local newspapers, casting a negative light on the TVA. Another, more likely possibility was that if the TVA used force consistently, it would have been viewed as a bully, creating fear and making residents less likely to cooperate. Often, the TVA allowed families more time to move, as long as their lives were not in danger due to rising floodwaters. Some families took this time to harvest additional crops or simply find another place to go.

Once the TVA purchased land and gave residents official notice of a deadline to leave, it became the responsibility of the caseworkers employed by the TVA's Family Removal Section to ensure that residents vacated their land. The importance of the caseworkers to the success of the TVA project cannot be overstated. They were the last in a line of TVA men (and women) to visit relocating families. They were the TVA representatives who arguably developed the closest, most personal relationship with the residents who had to leave.

Family Removal Section

While the Land Acquisition Division handled the TVA's needs with regard to land, the Family Removal Section dealt with what the TVA needed from the people who lived on the land. This division employed caseworkers like Martha Branscombe who were tasked with efficient and organized relocation of families before their land was flooded. The TVA hoped caseworkers would be able to aid families in acclimating to their new land and homes, leaving the families in as good, if not better, shape than they found them.[37] Family Removal Section staff represented the TVA to the residents most immediately affected by the agency's presence in the valley. For the residents in the Wheeler, Guntersville, and Pickwick reservoir areas who were being evicted from their homes and land, these were the individuals who forced them to confront a harsh reality: it was time to leave.

Caseworkers were crucial to the removal process. Bridge removals, road closings, and the removal of buildings were prohibited until the Family Removal Section gave final clearance, confirming that all residents were gone from the danger zones. Caseworkers were tasked to visit each family in the reservoir areas "as often as is thought wise and until removal actually takes place"[38] to ensure that families did move. Caseworkers closed a case when a family vacated TVA property, and then they assessed whether removal was successful based on the residents' new proximity to schools, churches, and stores; the quality of new housing; and their ability to secure employment in the new location. However, many cases were deemed closed despite lingering social problems, legal issues, or other difficulties that had not been fully resolved when the family left. Regardless of whether the family was better or worse off than before relocation, caseworkers often stopped assisting families after they left their property. And a relocation deemed "successful" did not necessarily mean the family was better off than before; the bar for improving conditions was exceptionally low.

Several criteria rendered a case unsatisfactorily relocated. For example, if a family knew they would have to move again because they had no sustainable crops or no farmland to work, their relocation was deemed unsuccessful. Some families were considered unsatisfactorily relocated if their land was not sufficient to meet a family's needs or if the land was of poorer quality than what they had forfeited to the TVA. Even so, relocations to poorer quality and/or less land were often considered successful if other factors in the family's life were improved in the relocation process. Families were also considered unsuccessfully relocated if it was obvious they would have trouble adjusting to their new area, such as if a black family had to relocate to an area that was mostly white.

It took a special type of person to work the frontlines of family removal, a very sensitive component of the TVA plan. Caseworkers' jobs were difficult and demanding. The Family Removal Section hired individuals who could compassionately deal with serious social issues and yet remain loyal to their job of relocating the family. They gave "careful consideration to the social and economic needs of persons who are removed and [made] every effort to help such persons avoid confusion and make proper adjustments."[39] This self-proclaimed "service" agency acknowledged the need to deal with the rural poor's many problems while facilitating the TVA's programs. Caseworkers worked on long-term improvement of economic and social conditions but also to deal directly with a number of immediate health, housing, and even "personality problems."[40] The concern voiced in the detailed narratives makes it clear that caseworkers cared about the families who were impacted by removal. They knew them and their lives on a deeply personal level. They saw the falling-apart homes, hungry children, and sick parents with their own eyes. They witnessed firsthand the heartbreaking poverty facing many north Alabama residents. Though they had a job to do, they did so with respect, care, and as much sensitivity as the job permitted. And they did what they could to ensure families were left in equal, ideally better, circumstances following relocation.

Their complicated and difficult jobs could sometimes also be dangerous, as they dealt with a good deal of uncertainty. Caseworkers traveled all across the Tennessee Valley by car and sometimes by foot (since many homes were not on paved or even passable roads), and they never quite knew what sort of situation they were walking into when they met with a new family for the first time. Their travels took them to extremely remote areas off highways and main roads, and into the backwoods and backcountry where few "outsiders" dared to travel. Their visits were unscheduled. Most residents were friendly or at least tolerant of the caseworkers' visits, but their unexpected visits occasionally caused intense negative reactions. Also, some families had negative views of the TVA, which sometimes translated to vocal opposition in the face of a caseworker. For example, L. B. Whitaker of the Guntersville reservoir area met a caseworker in his driveway with a shotgun, which he used to threaten the caseworker off his land.[41]

The TVA offered a vivid job description for ideal caseworkers in north Alabama. Caseworkers were to be "fairly mature" men who were 30–50 years old, had a "good education," were open minded and "without prejudices," held some knowledge of farming or life on a farm, "practical and with a good fund of common sense," and, most importantly, the "ability to meet . . . people . . . and gain their *complete confidence*"[42] (emphasis theirs). Those who had any experience with social services, farming, rural life, appraisals or land trading "of the honest, reliable kind" were considered good candidates.[43]

Given regional prejudices and the era, the TVA likely expected all of the caseworkers they hired would be white men. However, they found that hiring Birdius Browne, the only black caseworker in the Wheeler reservoir section, was helpful in working with black residents across the Tennessee Valley.[44] Martha Branscombe was also an unexpected hire; she was given the title of principal social caseworker and a salary of $2,600, only $300 less than that of A. L. Snell, the chief social caseworker.[45] They were required to work quickly, under what must have seemed like emergency circumstances, because residents officially were given only 2 months to relocate. Of family removal in the Wheeler area, one TVA document noted, "this reservoir area presented problems peculiar to itself and that varied widely from problems in other reservoir areas," due in part to sensitive race issues, intense poverty, lack of industries, and difficult terrain.[46]

Caseworkers took notes that were later transcribed into meticulously detailed narrative records for each relocated family. They collected basic demographic data, such as the names, ages, and occupations of each person in the home; relevant employment history; race; military status; income; possessions; type of house; descriptions of their surroundings, health problems, demeanor, and personality; and, in some cases, physical descriptions. Often, caseworkers noted in the narrative whether the family had a positive or negative attitude toward the TVA. The TVA engaged in this level of record keeping perhaps in an effort to verify to the public that the families who relocated were socially well adjusted after removal,[47] but caseworkers and the Family Removal Section ceased to keep records after they designated a family's case "closed." Sometimes caseworkers indicated they needed further relocation services,[48] but no record was created of whether those services were provided.[49]

Families who moved before the caseworker had the chance to visit their home had small files and very little information preserved. More complicated cases warranted creation of more in-depth files of one hundred pages or more. Caseworkers provided in-depth investigations of families for whom the removal process presented extra challenges, offering an understanding of the locals' way of life and enabling future research into whether their lives improved with the TVA's presence.

The TVA did not claim responsibility for finding displaced families new places to live; that responsibility rested upon the removed individuals. Instead, caseworkers reminded them of the necessity of moving and visited them as many times as they needed to encourage relocation. Family Removal Section orientation guidelines clearly stated that "the TVA is not empowered to give direct benefits to reservoir families,"[50] which was a reminder to caseworkers they could not provide assistance to families on behalf of the TVA. If any family demonstrated a special need, as so many did throughout the poverty-stricken lands between Pickwick and Guntersville, the family was referred to another service or relief agency. Evidence of referrals to other federal-level agencies can be seen in the final removal report for Wheeler; a total of 74 families were assisted by the Department of Public Welfare (27), Works Progress Administration (20), Rural Resettlement Administration (11), Cumberland Mountain Farms (3), and others (13).[51]

Despite the obvious sanitation, health, and social issues in this and many other homes throughout north Alabama, the TVA's caseworkers had not been hired to solve problems for north Alabama residents. Instead, they were tasked with ensuring residents moved quickly and quietly. They did have the ability to enact emergency removal by boat, but generally, the idea was for families to figure out relocation on their own.[52] Though the caseworkers' primary responsibility was to make sure people left the land on which they once lived, they were not heartless, unkind, or apathetic to the condition of the residents along the Tennessee River. Caseworkers often found medical care for residents and performed duties above and beyond what was expected of them. And in cases of extreme emergency, the caseworkers could arrange for a TVA-provided tent as temporary shelter, which on occasion proved to be better housing for the family than they previously had.

Some families saw the forced removal and relocation as a positive change. Russell and Maud Rogers and their five children, Jackson County residents, looked forward to relocation. When Mr. New, a TVA caseworker, visited on July 12, 1937, to check on their relocation status, he noted that their four-room house in "bad repair" was "meagerly but adequately furnished with cheap, worn furniture." Maud, who was pregnant and "engaged in fanning off flies" besetting their 19-month old baby, considered relocation a potentially positive thing. According to her, the family "could possibly better themselves by relocation."[53] Maud said she was "tired of sharecropping and having the family pay such a large sum of their crop to the landlord." Others felt equally as optimistic about relocating despite the fact that they had no choice in the matter. Relocating meant families might end up closer to schools, the center of town, churches, or medical care. It was unlikely that every aspect of a person's life would improve after relocation; a family might end up with a "mixed bag." For example, a family may have been "better situated" because of better access to a church or school, but not as well off in terms of the amount of land they were left to work. Given the extent of social problems in the north Alabama reservoir areas, there was substantial room for improvement, and the caseworkers took any small victory as a success.

People who later reminisced about relocation often had similarly positive attitudes. Two different branches of the Conner family in the Guntersville area were optimistic about relocating, even though Paul Conner admitted the decision about where to go after the TVA bought their land was difficult. T. L. Conner and his sister, Bobbie Conner Curry, the last person to be born on Conner Island before the TVA took ownership, remembered being excited about the move. For their parents, T. L. remembered it "was no big problem, I don't think. . . . We hadn't ever been nowhere different, and we got to see different country."[54] Bobbie noted the excitement they felt whenever they got to do anything different from their routine life on the farm. She said it was an exciting experience to spend the day with an aunt and uncle. Sometimes they'd "hitch up and borrow a 2-wheeled contraption and the horse would pull us, and that was sort of a nice experience." Paul's father, who was always interested in learning about ways to improve his farming methods, embraced a "pioneer spirit" and took the move as an opportunity to grow better crops and make more money. According to Paul, "I don't think it ever bothered my mother and father that much that, it didn't upset 'em that much because they knew the dam was going to generate electricity. Someday they'd have electricity. And we'd have the money."[55] Maxine Black's father, Thomas Williamson, comforted Maxine, who was sad about their land flooding, telling her "don't worry about it, everything's gonna work out alright." He resolved to work with the TVA to the best of his ability, learning how to improve his farming methods and grow better crops. Paul Conner's family felt it "was somewhat traumatic, the move itself, but [his parents] were not bitterly against it. They were not ticked off because dad loved FDR. And it supposedly would be a big improvement."[56]

It is unsurprising that residents who worked for the TVA clearing land said they had a positive attitude about the agency, despite facing removal. It makes sense that those who received a paycheck from the TVA were likely to say they were pleased with the program. It is also possible that the workers were fearful of losing their jobs if they spoke out against their employer to the caseworkers coming to discuss relocation. Still, as the TVA provided a higher income than many residents had ever seen, that was reason enough to feel good about the TVA.

Finding Out About Removal

North Alabama residents who needed to relocate found out about their fate first through hearsay, before they received formal letters from the TVA. Word about impending moves traveled quickly among communities. Hazel Moore Thompson, 8 years old when her family had to leave their farm in the Guntersville area, remembers hearing about having to relocate from other people in her community, as her neighbors vaguely "started saying we had to move."[57] Luther Tidwell, who was 11 at the time of his family's relocation, shared a similar experience. He remembers the day the family learned they would be moving: "My daddy come in and he looked like he'd been whipped. He said, 'Myrt [Luther's mother], we gon' have to move, they gonna take the property.' And everybody started crying."[58] Disappointment, fear, and hurt feelings were common initial reactions for families hearing the news.

Official word of a family's need to relocate came directly from the TVA, in a form letter mailed to each family. These letters served only to notify residents that their TVA did not address specifics of the large-scale planning efforts in these short communications. Residents in the Guntersville reservoir area received letters dated October 25, 1938, requesting that they move by December 31, 1938. The letters explained very clearly, "This advance notice is given to you in order that you may take advantage of the favorable fall weather to remove crops, buildings, and other improvements, as well as your family and livestock, from all TVA lands. We believe your cooperation in removing your interests earlier than December 31, 1938, will be of mutual benefit to both you and the Tennessee Valley Authority."[59] Follow-up letters were sometimes mailed, such as the one Roy Day in the Wheeler reservoir area received, reminding him he was trespassing if he did not leave immediately upon receipt of the letter.[60] Residents who stayed past the initial 2-month grace period were visited more often by caseworkers and sent written communication, like the letter, to encourage relocation. Fred Terry received a letter from the TVA stating "we've been lenient" and reminding him to move fast.[61]

Eventually, caseworkers' visits became memorable for relocating families. T. L. Conner recalled visiting TVA representatives confirming their need to move, as he said he "remember[ed] 'em coming around, appraisin' the land, and they'd take my father shopping for land . . . somewhere else."[62] Luther Tidwell remembered "an A-model Ford car came around the little mountain and up to the house 2 or 3 times, and I remember 'em saying they's from the TVA. And I didn't know what the TVA was, [but] undoubtedly they must've come several times."[63] Bill Hardin also remembered TVA representatives coming to his home. He said, "They came to the house and told us that they was going to purchase all the land, or condemn it, and we had no choice, we had to get out. And where they's going to back the water up, there wasn't any other connecting roads, so it was a matter of, you know, you had to get out [or be trapped in an area with no way out]."[64]

Valley residents affected by relocation responded to the news in different ways. Some did not believe they would have to move; those who owned their land believed it was theirs to keep indefinitely. Those residents quickly found out that if the government wanted their land, they could get it. Others did not believe that the water would flood as much of the land as the "TVA men" said it would. Maxine Black was one such resident. As a child who actively helped her father on the large farm he owned, she could not imagine that the water adjacent to the farm would flood the land on which her cattle and tall corn stood in the fall of 1936. She remembered, "My daddy would put a stake down [on their land], and he'd say, 'well [the water will] probably come to right here.' And we'd get on our horses in a few days and ride back down there. Well that water'd come to that stake. So he'd take it up, move it back so many feet . . . 'til it got completely filled and then it looked like an ocean compared to what it was. I couldn't believe it."[65] Her neighbors felt the same way: "Everybody couldn't believe it would happen. . . . [we] just couldn't believe that water would come out in these hollers."

The process of removal was different for each family depending on the amount of possessions they owned, their access to motor vehicles, and the distance they were moving from their home. Though most residents did not move very far and did not have many worldly possessions, they understandably wanted to keep what little they had. Some families carefully dismantled barns and houses to reconstruct elsewhere. That made it important to take care with moving, to avoid damaging the materials that might construct a new barn or home. George Hodge's father, Marion, converted a 1932 Chevrolet into a flatbed truck that he used to help people relocate from Guntersville to the Holland, Tennessee, area.[66] Those who had access to a truck were fortunate, but were the exception rather than the rule. Bill Hardin said moving was pretty much the same for all farm families, based on his experience: "Mules and wagons. That's the way we moved. Loaded up in wagons, mules, and the old road that's coming out of that Hambrick Holler came along the bluff line about where the dam is, and came on right at the base of the mountain, it's only one way in and one way out there, and other than trails. And that's where we had to move everything out."[67] It took his family a couple of weeks to move. They would "move a load out, like this weekend and . . . you know a day or so later we moved another load out. And the last trip I remember, my mother and my sister and myself and my older brother, we had to bring some of the animals out with us. We walked 'em up the mountain!"

Most residents left their homes within three or four visits from a case-worker. Depending on the number of families in the caseworker's area and the geographic area in which he or she worked, visits occurred any-where from daily to weekly. Some families still refused to leave after 10, 15, and even 20 visits. Families offered a variety of reasons for resisting their moves. Some offered the believable claim of running into difficulty finding a new place to live, some were too ill to move and had to wait to gain back strength, and some were simply not persuaded. W. G. Cagle, age 80, at first appeared to be willing to cooperate with relocation. However, he had still not left from land in the Guntersville reservoir area by March 14, 1939, and the deadline to leave was quickly approaching. He told the caseworker that he could "move when he well pleased."[68] The caseworker threatened to call in federal officers to relocate him; Cagle moved the next day. J. W. Floyd of Marshall County said he was going to "take his time" moving because the water would not cover his house and he had no money left with which to move. His caseworker had to convince him otherwise, eventually coercing him into leaving because he was on government property.[69] Those who did not move quickly enough risked having their land condemned; such was the case for Peter Roscoe Ivy.[70] The TVA had the legal authority to evict people from their homes, though it was in the agency's best interest to treat resistors with care, as peaceful relocation was crucial to maintaining their carefully crafted public image.

There were very few instances in which legal assistance was required for the north Alabama relocation effort. For the most part, caseworkers suc-cessfully persuaded people to leave peacefully, regardless of how they felt about it. That means they did their jobs for the TVA quite well. Neverthe-less, a few people threatened the TVA with either personal or legal force. Only one instance noted in the case files of a relocated individual hiring a lawyer was in the case of George Neville. His Decatur lawyer sent a scathing letter to the TVA, which ultimately did no good.[71] Mr. Neville, like every other resident on TVA land, had to leave.

In addition to receiving a form letter from the TVA and visits from case-workers, affected residents learned about the TVA from local newspapers. News coverage of family relocations focused mostly on the positive returns promised by the TVA, rather than on hardships experienced by families who had to move. The *Guntersville Advertiser & Democrat* and the *Limestone Democrat* ran editorials pointing out that "few people have stopped to think of what the building of Wheeler Dam means to Limestone County," but that the land "taken out of cultivation and off the tax rolls" would also result in a "protective belt around the waterfront to keep a timberline to prevent erosion and to keep out undesirable buildings and enterprises on the lake."[72] Newspapers admitted that some valuable sites would be submerged, such as the Harris School, but "TVA paid $4,355 for that privilege—a good bargain for the state," especially because the money would go toward building a new school, and the TVA allowed the existing school to remain in operation until the land was scheduled to be flooded.[73] Newspapers repeatedly challenged their own mild criticisms of the TVA.

In Limestone County alone, the TVA purchased 52,000 acres of land, which translated to one-seventh of the total acreage of land in the county.[74] The *Alabama Courier* considered the taking of some land a good thing; they wrote, "when timber from river lands are cleared away, the men who seek seclusion of the river bottoms for their location of booze factories will be greatly lessened."[75] The *Limestone Democrat* reminded readers that though this land would be "taken out of cultivation," the "million and one-half dollars to be paid for the fifty-odd thousand acres to be taken over by the TVA will be available for investment" in other things.[76] Initially, newspapers exaggerated the amount to be spent on land acquisition, with one early article claiming the TVA would spend "at least $3,600,000 in Limestone."[77] Wheeler's construction was "not an unmixed blessing," as the TVA claimed land that would never be replaced. But, ultimately, the *Democrat* assured residents, the TVA's land acquisition was "not harmful" to Limestone County because the TVA mostly purchased river bottom land that was of low tax value.[78] Tax values were mentioned in the *Huntsville Times* as well, which reported that the sale of 10,617.87 acres of land in Madison County at a total cost of $386,525.30 meant a substantial reduction in the county's taxes, but that deficit was compensated by the amount spent on land, which was "a godsend to many of the property owners."[79] The money the TVA paid farmers, they said, would be eventually invested in other areas, benefiting the economy.

The *Democrat* estimated the tax value of the lands purchased by the TVA at $13 per acre, which "means that over $675,000 of tax value goes off Limestone's books forever, entailing an annual shrinkage in state, county and school taxes." Limestone County later reported they expected they would quickly recover from the loss of $13,500 in taxes from the submerged lands in a few years by "the great increase in values by reason of TVA activities in this district." The semicritical editorial closed with this thought: "The Lord isn't making any more land, but He is permitting the birth of more people every day and they are going to need the land that's already here."[80]

The *Decatur Daily* also justified and supported land purchases in Morgan County, repeating A. E. Morgan's belief that "if a man is handling his land in a way that will destroy it, the part he cannot take care of should be taken away from him and given to someone who will farm it properly. A man has no natural right to inherit good land and pass on a waste to those who come after him."[81] The *Alabama Courier* reassured readers that the TVA was fair in their dealings with the people of the Norris reservoir area, implying they would be similarly fair in dealing with any Limestone County families who might be affected by relocation, and that the TVA should be given credit for "attempting to relocate and reestablish these families being moved out by flood waters in situations that will afford them equal if not greater contentment than they now enjoy."[82] Two weeks later, the *Courier* again reassured readers that those who were being required to sell land were treated fairly, and that owners "would profit by accepting the reasonable prices offered. . . . When a price averaging twenty-five to forty dollars is offered per acre . . . a man would do well to sign on the dotted line."[83] In this article, it's clear that newspapers were not only reminding residents just how fairly the TVA was treating everyone in the Tennessee Valley, but that people would be wrong to refuse to cooperate with the TVA. The *Guntersville Advertiser & Democrat* speculated about where the Guntersville Dam would be constructed long before any formal announcement confirmed its location in Marshall County. In an editorial, the *Democrat* wrote that if the dam was created "below Guntersville, we will not lose anything but some low land that is not being used any way, and of course we will lose some good farming land if the dam is located either below or above" the site. They hoped, however, that "land owners will be well paid for moving." The editorial went on to warn residents, saying "'[if you] have property here, it will pay you to hush until you can sell out, and those of us that don't have any property should hush, for we might discourage others. . . . We would starve to death. . . . A dam would make us like Chicago is to the west. Let's boost Guntersville! All you can say won't change the location of the dam."[84] Interestingly, Guntersville newspapers did not report on the extent to which TVA land acquisition changed the topography and existing structures in place in the town of Guntersville. Nor did Guntersville newspapers report on the hardships faced by those being relocated. Editorials and articles largely encouraged readers to focus on the positive aspects of the TVA's presence rather than think critically about the government agency as a potential threat to their livelihoods.

Newspapers did not address the fact that landowners fared much better financially than the tenant farmers, sharecroppers, and squatters who also were forced to leave. Very few residents in the reservoir areas actually owned their land. Only 61 out of the 842 in Wheeler were landowners, 152 out of 1,182 were owners in Guntersville, and 126 out of 499 were owners in Pickwick,[85] so it's safe to say that most people who had to relocate from TVA-purchased land did not receive any financial compensation for their move. Owners generally tried to keep their tenants with them as they moved to other farms, but with the land shortage, many farmers had no choice but to downsize their operations, which meant less land for the owners and even less land for tenants and sharecroppers.

The promise of new jobs and rural electricity, the supposed end of the sharecropper system, and increased education about better farming practices were all aspects of the TVA that seemed exciting to many in north Alabama, even if dam construction resulted in temporary hardships. However, not everyone agreed. The TVA's promises fit into a larger narrative constructed by another TVA division far removed from the frontlines of relocation, one that breathed life into the utopian promises made by FDR. The Information Office, the TVA's public relations arm, capitalized on these sentiments and embarked on a plan to shape public opinion. Much like the caseworkers sent to appraise land and assist relocating families, the Information Office sought to emphasize the promises of the TVA over any potential or real problems facing the residents. An investigation into how the Information Office functioned in the 1930s, and how the TVA deliberately worked to convince journalists and other opinion leaders to spread the TVA gospel, yields insight into the power of media messages in the creation of public opinion.

4

Selling the TVA

Knoxville resident Charles Krutch stumbled into photography using a hand-me-down Kodak camera. Borrowing the camera from two friends who had recently had a baby, he promised he'd learn how to use it to take professional quality pictures of their baby. Little did he know that camera would be his gateway to a life-changing opportunity, or that a baby would not be his first test subject.

Shortly after acquiring the camera, and still unsure of how exactly to use it, he found himself at a cocktail party, honoring new Knoxville resident and TVA board of directors member David Lilienthal. Their chance meeting changed the course of Krutch's life and TVA history. The two men started talking. When Lilienthal asked Krutch what he did for a living, Krutch innocently responded that he was an amateur photographer, and he hoped his pictures would capture more than just facts: he wanted to "catch the meaning of a picture." Lilienthal enthusiastically responded, "You're just what the TVA needs. Why not be a roving photographer with no specific assignment?" Having no other viable employment options at the time, Krutch went the next day to sign up as a TVA employee, traveling across the Tennessee Valley to take pictures of dams and the people they were meant to serve. Krutch ended up staying with the TVA for 20 years pursuing a career that resulted in international acclaim for his artistic photographs, mostly of TVA projects.[1]

Though TVA engineers had been extensively documenting all aspects of dam construction in photographs from the start of those projects, until Krutch was hired, there were very few pictures suitable for public distribution—pictures that would explain, better than words ever could, why the TVA was needed. The public needed to see struggling farmers, eroded landscapes, and the sheer massiveness of the dams to understand fully what the TVA meant to rural southerners. Krutch's pictures, arguably more than the words that accompanied them in print materials, captured the essence of life in the Tennessee Valley and were extensively used as publicity materials to convince people to support the TVA.[2] His pictures were so artistic, a TVA chief engineer once remarked that the TVA did not build dams that looked like the ones in Krutch's photographs, implying that they made the dams look too beautiful.[3] Krutch's place in the Information Office, particularly in his role as chief of the graphic arts section, became iconic, as would many of his pictures.

Krutch learned quickly how to take stunning pictures, and he eventually earned widespread recognition for his photographs, many of which were displayed in exhibits around the world, including one in New York City's Museum of Modern Art. In fact, his photographs are still used in TVA publicity materials today.[4] But he did more than simply take good pictures to accompany media stories about the TVA. He smartly, and quickly, figured out how to multiply the number of eyes that saw those pictures. He once described the "trickery" he used to get his best shots widely distributed for free: he made sure to "give [the picture] to one of the reporters on one of the local papers [and then] sell it to the Associated Press for five dollars. They would send it out all over the country on the wire, and it would be TVA propaganda."[5] In addition to knowing how to take amazing pictures, Krutch also knew how to market his work and the TVA.

Krutch was one of three men at the helm of the TVA Information Office. His influence on media messages communicated by and about the TVA is undeniable. Unconventional strategies like the ones he used to distribute his photographs were characteristic of the way the TVA carried out most of its projects. As a new agency that acted like a business but had the backing of the government, the TVA's earliest employees had the luxury of figuring out how to best do their jobs while keeping the ultimate goal of making the TVA a success front and center.

But that success depended largely on the work done by the Information Office. In order for the TVA, in many ways an experiment, to work, they had to sell electricity in north Alabama. And to sell electricity, they had to get potential customers to understand not only what electricity was, but how it could be used. They also needed to develop TVA loyalists who would help shape public opinion in the coming years. In short, the TVA needed north Alabamians to feel good about the TVA and what they were doing with their lands and people. Joseph Swidler, an early lawyer for the TVA, summarized this challenge, which essentially was a public relations challenge: "The TVA had really no police power. . . . It could only be successful by convincing people of the Tennessee Valley that it could help them and that the program was sound."[6] If it could not convince the people of north

Alabama to use the new, cheaper hydroelectric power, the TVA program would surely fail. Though north Mississippi was also important in the early days of the TVA, the agency would be serving many more customers across north Alabama. This meant that local residents, many of whom were wary of electricity or what it meant to their way of life, had to first be educated about the TVA and had to see it as a positive force for good. It also meant the rural poor who could barely afford food would need to soon start purchasing electricity, something that, while desired, may have been viewed as an unaffordable expense, no matter how cheap the rates TVA offered.

To help the public see the good promised by the TVA program, the Information Office created a multiplatform, comprehensive communication plan designed to combat public criticism and emphasize the TVA's successes. Most news, information, and updates about the TVA were delivered to north Alabama residents through the widely used media channels available in the era—radio, newsreels, and print media. While media messages did not reach audiences as quickly or as large as those of today, word-of-mouth helped to spread the word about what the TVA was and what it potentially meant to the Tennessee Valley. At a time when just about any potential plan for economic development was well received, the TVA enjoyed an initial positive public response. The promise of rural electricity and jobs was exciting to many. But TVA had its share of critics, many of whom were quite vocal and would continue to be critical throughout the most crucial time of the TVA's early development. A concerted public relations effort was needed to keep attitudes positive, shape public opinion, and counter criticism. It was crucial for the TVA to capitalize on the excitement many felt toward the government agency that promised to bring the South out of poverty. The TVA needed to make sure that criticism was kept to a minimum while praise was prolific.

The Information Office was responsible for all external and internal communications for the TVA. Similar in structure and function to a modern-day corporation's public relations division, the employees of the Information Office almost immediately went to work to sell the TVA to Tennessee Valley residents. The name of the office was carefully crafted to deflect criticism. The term "public information" office, as opposed to the "publicity" or "public relations" office, was chosen because the latter terms had the potential to "arouse a certain antagonism or defensive attitude in the minds of editors and others whose cooperation will frequently be sought."[7] From then on, the Information Office was careful to avoid labeling themselves as the public relations arm of the TVA. Instead, their presumed focus on "publicity" framed their efforts as simply that: publicizing the TVA to the South, the United States, and the world. One historian noted that the amount of publicity about the TVA "offended nobody," as it tended to emphasize the TVA's role in comprehensive development of the Tennessee Valley.[8]

The first director of the Information Office was William L. Sturdevant, who built his career as a journalist and newspaper editor before joining the TVA. A graduate of Alleghany College, he was serving as the editor of the *Louisville Herald-Post* when he joined TVA, taking a pay cut for the new position. He'd previously worked on the Youngstown, Ohio, *Telegraph*, the New York *Telegram*, and the *Birmingham Post-Herald*. Assistant director Maurice Henle also had extensive newspaper experience, previously writing for *The Cincinnati Post*, the *New York Daily News*, and the *St. Louis Times*.[9] The decision to hire two newspapermen to lead the publicity effort was intentional. Print media would be the main avenue for getting information out to the public, and both Sturdevant and Henle had the experience, skill, and professional networks to give the TVA a solid start at generating positive print publicity. Together with Krutch, the three men led the Information Office until the early 1950s, when Sturdevant and Krutch both retired.

Other employees were essential to the office's functioning. Staff members focused on generating written communication, specializing in topics related to agriculture, rural electricity, engineering, and general TVA information. These staffers frequently accompanied journalists and editors who visited TVA projects to do research and to write about the TVA for other publications. One writer was responsible for captioning all pictures that were sent to publications or included in pamphlets. Field representatives worked in Chattanooga, Muscle Shoals, and Washington, DC, offices during the Pickwick, Wheeler, and Guntersville reservoir construction. The remaining office staff worked diligently to respond to letters and requests for information, write press releases, and carry out other daily tasks. Their staff grew from eight to forty-four during the period 1932–1939. In the first 5 years of the TVA's existence, their budget was $757,237.22—a considerable sum, large enough to sponsor widespread publicity efforts.[10] The rapid growth of the office staff and generous budget reflected the Information Office's importance to the overall TVA program.

The Information Office had unique and crucial responsibilities. It acted as an intermediary between the board of directors and the public, ensuring that TVA officials were aware of public sentiment so they could craft public messages accordingly.[11] Sturdevant routinely attended board meetings starting in 1935 to further solidify the connection between the Information Office and TVA leaders.[12] Their strategy was to sell the TVA to their stakeholders in the Tennessee Valley by working with media outlets to promote their objectives, and to combat public criticism in the same way. The

Information Office maintained a technical and research library; prepared the TVA annual report for Congress; created displays for state and world's fairs; installed permanent promotional exhibits at each of the dam sites; wrote speeches for board members and other TVA officials; prepared press releases distributed to an extensive list of newspapers and opinion leaders; created informational pamphlets; distributed movies about the TVA to schools, businesses, and theaters across the country; and provided written responses to all requests for information.

Information Office staff also maintained an index of newspaper and magazine articles written about the TVA. Through this project, the Information Office kept abreast of the most current public opinions about the TVA, allowing them to adjust future messages to align with what they knew the public saw as positive or combat public criticism. Articles in praise of the TVA were many, but the Information Office also kept files on "all hostile propaganda in press or elsewhere."[13] The index was important as the TVA answered smear campaigns launched by private power companies and combated a heated legal battle against them. The record of writings about all aspects of the TVA allowed staff writers to respond with their own campaigns.

In today's media-saturated world, it is difficult to imagine launching a publicity campaign without the help of television, social media, and other digital marketing tools. The Information Office staff creatively communicated about the TVA to local, regional, national, and international audiences using a range of what would now be considered low-tech media. The Information Office heavily emphasized printed messages, with one self-imposed limitation: as a government agency, they opted not to advertise in the traditional sense of the word.[14] Careful to avoid accusations of spreading TVA propaganda, they incorporated a large number of grassroots, interpersonal approaches, rather than authoritative ones, to sell the TVA.[15]

The Information Office "ignored" what they saw as the three prevailing public relations strategies of that era. One strategy was to "draw a cloak of silence" around their happenings, which would result in no publicity at all. A second dominant strategy stemming from the belief that any publicity is good publicity was also avoided. (The TVA recognized the benefits of being strategic about what media messages existed about the agency.) The third strategy was what they saw as the usual approach in government agencies: "stereotyped hand-outs and . . . rigidly controlled censorship."[16] The TVA certainly was a unique force in American government that required a delicate approach to selling its program for rural electricity and regional improvements. As a result, decisionmakers in the Information Office developed an innovative, unique approach to public relations, one that would ideally result in their target publics' complete support of their goals.[17]

The professional backgrounds of the three men in charge of the Information Office—William Sturdevant, Maurice Henle, and Charles Krutch—greatly influenced the philosophy and strategy of the TVA public relations effort. Even though Lilienthal was not formally part of the Information Office staff, he had many professional and political relationships that proved useful to them.[18] He also shared his ideas for creating a positive public image for the TVA. These included a TVA license plate similar to the one that promoted Boulder Dam. He felt "a little of the right kind of publicity . . . inviting the people . . . on the road to 'See Norris Dam'"[19] would help bolster the TVA's visibility throughout the country. Despite several of Lilienthal's suggestions being largely ignored by Sturdevant and the Information Office, the interpersonal and mediated approaches that were used during the beginning of the office's existence were successful, and they remained the primary ways the TVA communicated about itself for several years. An emphasis on print media, press releases that would become newspaper and magazine articles, and award-winning informational pamphlets formed the foundation of their operations.[20]

Mary Utopia Rothrock, an early librarian for the TVA, noted that "the only way regional growth or any sort of social development comes about . . . [is that] people have to want it. You can't impose it on them. If you impose it on them, it is just exactly like putting a snow-covered long log on that fire. It will just fail right away."[21] Apparently well aware of this notion, Sturdevant and the Information Office staff designed what they considered a new strategy that emphasized the positive aspects of their organization, balancing media saturation with a grassroots interpersonal communication strategy. Though they kept their messages, as they called them, "factual," reporting on things like construction efforts and the adoption of rural electricity, the underlying message in most communications was that the TVA was fulfilling its promises to the Tennessee Valley. The public relations machine at the Information Office constantly reminded Tennessee Valley residents and the rest of the world that not only were TVA programs working, but they were the right thing for the South. This overarching message was reflected in nearly every pamphlet, speech, letter, and other communication distributed by the office.

Early employee Edward Falck noted that the massive amount of TVA publicity was exciting. Information Office workers had a ready audience of people who sincerely wanted to know what was happening with the TVA, and they responded with interpersonal strategies that included writing letters and making phone calls, in addition to preparing speeches and crafting articles that were published in magazines with national audiences. Best of all, the publicity was largely affordable, with Information Office staffers crafting one message that was amplified over several platforms with little additional cost. For example, many of the speeches Lilienthal and other TVA leaders delivered were replicated in newspapers or broadcast on the radio for free.[22] Liberal and conservative presses alike were ready to share information about the TVA; magazines "from the *New Republic* to the *Saturday Evening Post*" featured TVA-related stories during the 1930s.[23] The comprehensive publicity plans created by the TVA's Information Office were creative and so widespread that they were arguably inescapable.

Interpersonal Strategies

The TVA Technical Library served as a repository for facts and figures about the TVA and its dams. A helpful resource for anyone interested in researching a multitude of aspects about the TVA, it did more than just allow the TVA to demonstrate that they were willing to be transparent about projects. TVA librarians readily assisted everyone from journalists to graduate students who came to learn. Speaking specifically about students coming to the library to research the TVA, Sturdevant once commented that "most of these students will be teachers. . . . After a year's study of TVA, it would be impossible for them to be anything but sympathetic about the TVA program."[24] The library went beyond functioning as an educational repository, serving also to recruit TVA loyalists across a number of academic and professional fields. The TVA also sought to capitalize on the expertise and enthusiasm of its many employees by encouraging them to contribute articles to trade publications and academic journals.[25] This strategy diversified the audiences who learned about the TVA.

TVA dams generated more than just hydroelectric power. Many were fascinated with these massive structures because no one in north Alabama or anyone in the South had never seen anything quite like them. The TVA soon recognized that the dams themselves could serve as a promotional arm of the agency. In fact, someone once mentioned to Lilienthal that "the best propaganda you have for your project are your dams. People go down there antagonistic and they see those jobs and come away ready to go out and fight for you."[26] Visitors came from all over the country to see the finished dams, and the Information Office celebrated the dams' importance, magnitude, and innovation in visitor center exhibits. Graphic arts staff prepared permanent installations at Pickwick, Wheeler, and Guntersville

Dams that showcased the amount of electricity generated and the regional improvements facilitated by dam construction. By 1938, just four years after the TVA's inception, an estimated 3.5 million people had visited a TVA project site. In north Alabama, Pickwick attracted 246,333 visitors, Wheeler had 176,704, and Guntersville saw 87,063 visitors from the time they opened to the public until 1938.[27] The TVA's dams made tourist destinations out of tiny towns across the South.

Understandably, local residents impacted by family relocation were just as curious about the dam as others, and they were intimately familiar with why the dams were needed, the types of improvements they offered, and the amount of electricity being generated. But visitor center exhibits paid little attention to relocated families. The number of families moved, the graves that were exhumed, and the continued social and economic problems still facing the poorest in the rural South were not part of the TVA's exhibits.

The Information Office also worked to share the TVA with people who did not have the opportunity to see one of the dams in person. Graphic Arts staff designed world-class exhibits for state and world's fairs, bringing the TVA to a much broader audience. Such exhibits regularly included a short film, a map of the Tennessee Valley, dioramas that mimicked the landscape of the Tennessee Valley, and pictures of farmers using electric engines and modernized farm equipment powered by the TVA. These exhibits featured the successful aspects of TVA programs, from rural electrification and farm improvement, to flood control, improved navigation, and modernization.[28]

The TVA recruited enthusiastic and talented workers, and the agency capitalized on the staff's early enthusiasm to cultivate a sense of loyalty among employees. A course in "The Meaning of TVA" was offered to employees as a way to fully socialize them into the agency.[29] Each dam site eventually became its own thriving community, bustling with skilled and specially trained workers. The TVA housed many employees in dormitories and small cottages near construction or onsite and provided cafeterias, recreation areas, and other common spaces. Agency-sponsored publications, such as the *Pickwick Papers*, for the Pickwick Dam community, the *Wagon Wheeler*, for the Wheeler Dam Community, and *Chips from the Reservoir*, created for those involved with reservoir clearing out of the Decatur, Alabama, office, helped keep worksite residents updated on local news and events and fostered a sense of community and a sense of purpose.[30] An additional motivation for the publications was to give teenage boys living in the community something to do with their free time, as they performed the typing, copying, layout, and art needed to bring the newspapers to life although the verbal content was generally contributed by the adult employees.[31] The TVA banked on the idea that motivated employees interacting with people outside the encampments would be more prone to speak highly of the TVA.

News, jokes, announcements, events, cartoons, historical material, and personal experience stories largely comprised the content of these employee publications, which were favorably received. The *Pickwick Papers*, coordinated and prepared by local high school students, was found by TVA administrators to do a considerable amount of good among TVA work camps.[32] Employees agreed that these publications were "eagerly received," "cement[ed] more closely the organization as a whole," "established a feeling of unity," and maintained "the morale of the division."[33] Though not directly related to the activities of the Information Office, the TVA board of directors and the Information Office were aware of their existence and came to recognize their importance in helping employees be good stewards of the TVA.

One of the most impressive and time-consuming projects undertaken by the Information Office staff was personally responding to every letter or written request for information received, whether those requests came from journalists, politicians, or citizens across the country. According to their estimates, from 1933 through mid-1938, the Information Office received on average 30,000 letters each year, more than 100 letters every business day. Every single letter received a response.[34] In providing that level of personal communication, the TVA felt it was creating "word of mouth sales agents."[35] Every carefully crafted letter sent back to a resident of the Tennessee Valley or beyond was an opportunity to highlight the positive aspects of TVA, minimize criticism, and counter arguments against their mission. Information Office officials readily mailed pamphlets to those who requested more detailed information, particularly in response to commonly asked questions and concerns, many about the new power program. For example, those new to electricity wondered about what the electric current meant for their homes and livelihoods, resulting in questions such as, "Can we use electricity for roasting coffee?"[36] The TVA thus engaged in a considerable amount of basic education regarding how residents could use electricity to improve their everyday lives.

The TVA also dealt with more serious issues, particularly from those who were forcibly relocated from their homes. Some wrote the TVA to request permission to stay where they were, for a lack of a place to move to next.[37] Some pleaded with the TVA in writing to send food for their starving children.[38] Caseworkers were made aware of requests coming from families in the reservoir areas so they could be addressed as much as possible during home visits.

Whether the letters came from children or congressmen, the Information Office staff sent a response. The time commitment involved with this process was commendable, but at the time, it was common practice among government agencies. The TVA, however, found a greater purpose in their letters program. In responding personally to every message, they earned great word-of-mouth support.[39] Edward Falck, a rate engineer for TVA, provided an example of how the TVA used its letter-writing program to further its goals. He spoke of how many farmers wrote to the TVA asking how they could get power for their farms. The TVA's response was often to encourage the farmer to organize a small group meeting with neighbors so TVA officials could come to make a presentation. Falck remembered that "the farmers would come with their wives, family, and many had children in arms, at night and we would have a two-hour meeting with them and we would tell them if they would band together and have not less than 100 families and form a cooperative, then we said we would work with them." These visits were gratifying to Falck because, he said, he felt he was able to provide a service to people who were anxious to reap the benefits of electricity on the farm. Families "who didn't have electric power to read by and to have their children read by. They were . . . willing to do anything possible [to get TVA power], including chopping down trees and digging post holes to put the poles in, and helping to string the wire—anything to get the juice."[40] Information Office staff made the most out of every request received, whether it was to respond with a personal note, drop an informational pamphlet into the mail, or meet with local farmers who wanted to learn how to use TVA power.

David Lilienthal engaged in his own personal letter-writing campaign as well. Ever concerned about his work as a public ambassador for the TVA, he realized the need to use concrete examples in the many speeches he'd deliver to audiences across the South. He once wrote to select users of TVA power across Mississippi, Tennessee, and north Alabama, saying, "I would like to find out what you think about TVA in your home. I can do my job a great deal better if I know just what has been your experience, and your neighbors', regarding TVA electricity."[41] Though it is unclear how he selected his recipients, it is apparent that he wrote the letter targeting a few specific groups, such as mayors, women, business owners, and school superintendents along with heads of households. The responses poured in, some written at the bottom of Lilienthal's original letter, some on lined paper, and some typed on business letterhead. Most praised the TVA. W. A. Graham, a principal in the Florence City School system, responded, "We have lights, radio, refrigerator and an iron, are going to get a (water) pump this month and hope to buy a stove in the early summer. Lights and radio have helped make the long winter more cheerful for my mother and sister. The pump will mean freedom from our slavery to the rope and pulley." J. H. Elliott of Ardmore, Alabama, said that electricity "means very much to my entire family. Everyone here in [Limestone] county is well pleased." J. H. West of Tuscumbia, Alabama, said he was "100% for TVA and I have heard no complaints from anyone." And M. J. Easter of Athens, Alabama said "TVA electricity has helped the people of Limestone County more than anything we have had for the past 20 years. . . . TVA has given people a new hope during one of the greatest depressions."[42] It should be noted that Lilienthal did not target relocated families—those who were disadvantaged by the TVA's land purchase efforts—in his letter-writing campaign. The people Lilienthal contacted were likely people who already had electricity supplied by Alabama Power, who saw their rates drastically slashed by the TVA's presence, clearly a group that would have had positive opinions.

Lilienthal incorporated the comments he received into a series of speeches he delivered across the Tennessee Valley, using the people's own words. As a result, even though he was largely an "outsider" of the South (as a former Wisconsinite), the people of the Tennessee Valley trusted and liked Lilienthal. He once commented on his outsider status: "I think the fact that I was a Yankee may have actually been a helpful thing rather than the reverse. Sometimes it is the cousin from a long way off that you are more likely to believe than the people you are closer to."[43] Lilienthal and others diligently worked to meet the people of the Tennessee Valley in person, especially those in north Alabama, through numerous speaking events in the earliest years of the TVA's existence.

The original board of directors, especially Lilienthal who was a charismatic orator, spoke extensively to large and small audiences across the country, championing the TVA's work. Each message was tailored for the audience, whether it was a group of TVA employees, university graduates at their commencement, or a civic organization such as the Rotary Club. And in the TVA's early years, transcripts of most board member speeches were sent out as press releases to local, regional, and national newspapers, or the speech might be broadcast simultaneously on radio stations. It was important, then, that speeches be carefully crafted to reinforce the TVA's mission and aligned with Information Office objectives. Lilienthal in particular realized that the people of the Tennessee Valley had repeatedly been written off as an economic problem, which he recognized had eventually led to demoralization and a perpetuation of their desperation. He felt Tennessee Valley residents needed a self-confidence boost, and this optimism was a theme in many of his speeches.[44]

Lilienthal championed the TVA's cause in every speech, reminding his audiences that not only would TVA programs benefit them, but they would be the ones to help bring positive change to the Tennessee Valley. Talking of the future of industry in the South, he claimed it would "be the scene of an expansion of industry which . . . will change the economic life of the South . . . [and] TVA made possible for this area to develop its natural resources."[45] He lauded the sale of TVA power[46] and held up the small north Alabama town of Athens as an example of the TVA's success with yardstick rates, noting that city revenues had increased even though electricity rates decreased from those put in place by Alabama Power.[47] Lilienthal spoke of the TVA's impact in Lauderdale County, Alabama. As a result of the TVA's presence, he said, this "cotton farming county" had transformed into a modern one: "168 [farm] customers have lights, 75% have radios, 48% bought large size refrigerators, 15 installed electric ranges."[48] He used statistics like these to show that the TVA was making a difference in the lives of farm families almost immediately.

It was common for speeches to refute popular criticism and reinforce loyalty to the TVA. Lilienthal once acknowledged his Wall Street critics and agreed that they should oppose the TVA because, if it was successful, "the economic trade balance will be improved in our favor, less draining off of assets of our area into the hands of the New York financial groups."[49] In a speech to TVA employees during the summer of 1936, Lilienthal stressed loyalty to the TVA, stoking the early enthusiasm many employees had for their jobs. He spoke of a higher purpose for TVA employees, noting "loyalty to this project, loyalty to our objectives, loyalty to the idea that the President had in his mind—there is a loyalty that we can all tie to and that will carry us over many, many difficult places. . . . TVA is about people and for people. TVA is about men and women and children."[50]

Lilienthal was not the only board member to make frequent speeches; both Harcourt Morgan and A. E. Morgan spoke on behalf of the agency, generally on topics representative of their respective duties on the board. However, the ways in which the speeches were received by the press and, essentially, the public, differed. For example, A. E.'s speeches focused on the social reforms afforded by TVA programs. Occasionally he offered incorrect information to audiences, or his messages were miscommunicated by the press; journalists covering Lilienthal's speeches, which focused on rural electrification and the public–private power debate, praised him and his messages while criticizing A. E.[51] This tension, played out in public forum, may have been one of the first indications of internal conflict among the TVA leadership—just the sort of scandal the Information Office would have hoped to avoid or minimize with public communication.

Frequent visits to the Tennessee Valley were not just for speeches. Often, there were celebrations of TVA such as jubilees, parades, picnics, and dinners.[52] Limestone County, Alabama, began celebrating the TVA as early as December 1934, when the local chamber of commerce hosted a dinner for thirty TVA staff members in appreciation of the positive impact the TVA had already had on the county.[53] Lilienthal returned the adoration expressed by local residents,[54] and newspapers continued to praise residents, especially in Athens, for welcoming the TVA with open arms.[55]

Although staged events such as these resulted in positive attention from newspapers, the relationship between communities and TVA promotions was not always smooth sailing. Nearby Hartselle was a harder sell than Decatur and Guntersville. Lilienthal once journaled, "in the sleepy, rudimentary little town of Hartselle, where I sat and talked to perhaps 20 men. . . . They looked at me like I was a three-headed calf, at first."[56] The reaction of Hartselle residents may have been similar to responses in other small towns and rural areas when Lilienthal or other TVA representatives visited. It's unlikely that rural residents undergoing forced relocation attended the celebrations, demonstrations, or movie screenings due to the difficulty they faced in traveling to city centers and their lack of disposable incomes; people whose walls were covered with newspapers, whose diets consisted of cornbread, and who consistently suffered from malaria lacked money for these luxuries.

Mediated Strategies

Interpersonal communication helped create TVA supporters in communities who would then tell others about the benefits of TVA. This personalized level of communication was a helpful strategy that extended the TVA's mission. However, it was not the only option the Information Office had in promoting the TVA. Media messages about the TVA heard on the radio, seen in movies, and read in magazines and newspapers dominated the media landscape in the rural South during the 1930s. More than any other government agency, the TVA heavily promoted itself using all available media outlets. Unlike other agencies specializing in relief, the TVA was not intended to be a temporary agency. To work effectively, it needed to develop staying power. Long-term public support, then, was necessary to combat criticism and promote the TVA across the South, the nation, and the world.

Maxine Williamson Black looked forward to Saturday evenings. Her family was one of the few in their Limestone County community who owned a radio. Neighbors came from near and far to listen to the Grand Ol' Opry, broadcast out of Nashville, and the occasional professional fight, on Saturday nights. For those like Maxine and her family who lived far away from the town center, the radio was the sole source of entertainment and a connection to the world beyond the pasture. The radio also reported important news, in a format more accessible than even the weekly newspapers that made their way to their home. It was on the radio they learned of impending floods—when they would need to help Knight's Island residents walk the cows over—and heard the latest updates on the TVA's plans to pull the South out of poverty. Radio use was common and growing throughout the country in the 1930s. People adored this new media for both entertainment and educational purposes. Despite the growing popularity and prevalence of radio, its influence in the South "lagged behind" the rest of the country in the late 1920s.[57] In Alabama, there were only twelve radio stations in the earliest days of TVA, and only three active stations in north Alabama.[58] Those who could afford a radio, like Maxine's family, tended to be hubs of social life on the farm.

The Information Office eventually worked with the TVA board of directors and other representatives to prepare content for radio programs. For example, the Huntsville station WBHP offered 30 minutes each week to county extension agents to talk about farming-related issues. The TVA pitched the idea of running a quiz show during one of these segments that would center on the TVA's erosion control work in north Alabama.[59] Most radio content related to the TVA, however, consisted of speeches by TVA officials, which were broadcast over the air.

Harcourt Morgan delivered a nationally aired address on September 26, 1934, providing the country with an update on the TVA's progress while reminding listeners of its importance. Speaking directly to farmers, he reminded listeners that there was still time to "save our land," and emphasized the TVA's commitment to helping farmers reap economic benefits.[60] A similar national radio address by David Lilienthal 2 years later emphasized that "130 million men, women and children own the TVA" and provided updates on the TVA's soil conservation and electricity programs.[61] Speaking before the Shelby County Young Democratic Club in Memphis, David Lilienthal's address was broadcast over WREC to a much larger audience. Elevating the TVA's importance, he said, "TVA is not merely part of the New Deal—it IS the new deal." Imagery of a life free from "no help wanted" signs, plentiful food and clothing and shelter, adequate housing, and better educational opportunities for children permeated his message, and these were tied directly into the TVA's overall mission and role in the Tennessee Valley.[62] By relying on radio broadcasts, messages like these were echoed across the South as TVA leaders emphasized its role in modernizing the region.

Hazel Moore Thompson, like Maxine Black, also enjoyed the Saturday night radio programs. But Saturdays were also special because, after she and her siblings finished their chores on the farm, their father gave them nickels to see the movies in town. Downtown Guntersville's movie theater offered times of relaxation, a respite from their hard-working farm lives. Occasionally, before the film, newsreels highlighted important stories of the day. Eventually, reels incorporated imagery of the progress the TVA was making across the South, including nearby Guntersville Dam, which they'd heard would bring electricity to even their small, rural farm house. The TVA was a common subject in nationwide newsreels of the day.

Several different film genres were popular at the time and thought to be potentially influential, including travelogues and documentaries. The TVA capitalized on the movie industry's extensive reach by creating their own promotional films. By June 30, 1938, more than half a million people across the country had watched one of seven TVA films, and an additional million people visited the TVA's visitor centers, where the films were shown.[63] Motion pictures offered more powerful storytelling than print or radio by incorporating moving pictures, still photographs, sound, and narratives. TVA films, like the popular *TVA at Work*, were distributed across the country to movie theaters and even schools, libraries, and community centers. In this black-and-white film, the TVA illustrated its earliest work at Norris, Wilson, Wheeler, and Pickwick Dams and showed footage of the beginning construction of Guntersville Dam. Audiences witnessed excavators digging deep trenches, men clearing land to be flooded, and the majestic, finished dams.

The film, silent except for the generally upbeat soundtrack performed by a band—perhaps the US Marine Corps Band, which performed music for at least one other TVA film—showcased scenes from Tennessee Valley life before and after the TVA's presence. It opened with a storm, including dramatic lightning and heavy sheets of rain, and its aftermath, a destructive flood in a rural area. Viewers are reminded, with descriptive text, that Congress created the TVA to prevent those scenes from happening, and showed on a map where Norris, Pickwick, Wilson, Wheeler, and Guntersville, the main channel dams, were already in place or under construction. Still images of Pickwick, still under construction at the time, were complemented with film footage taken as a car drove toward Wilson Dam from a rural country road. Wheeler's massive locks are shown opening for a barge to go through, illustrating the ways the dam improved river trade in a previously impassable area of the river. Throughout the film, audiences see men at work on TVA construction and land-clearing projects and productive farmers in the field, and inside the furnace at Wilson, manufacturing fertilizer. Viewers get the sense that the TVA was responsible for these jobs in a previously jobless area.

Next, the film focuses on what the TVA has meant to farmers, specifically describing the importance of nitrates produced at Wilson Dam and the relationship of farmers to extension agents who were helping them learn new soil maintenance and preservation techniques. Audiences are shown a side-by-side comparison of two plots of land planted at the same time: one, treated with fertilizer, which was lush and productive, and the next, with no fertilizer, dry and barren. Further images of severe erosion were shown, illustrating the need for the TVA's commitment to helping farmers prevent more soil loss. Farmers are shown spreading fertilizer with mule-drawn carriages, and the film describes farmers cooperating with each other and extension agents to continue using these techniques.

Lastly, the issue of rural electrification is addressed. Men are shown unloading an electric stove out of the back of a pickup truck, delivering it to a farm wife standing by to watch them. Tall, imposing power lines crossing the countryside are shown along with an electrician climbing a pole up to the top to perform maintenance. In closing, viewers are reminded again of the destructive floods that wrecked the Tennessee Valley before the TVA's presence, and then the improved, productive way of life, modernized with electricity. The final image is that of a glowing light bulb, symbolic of electricity and the power the TVA represents in north Alabama and the entire southern region.

Both Limestone County's Ritz Theater and Guntersville's Palace Theater held free showings of TVA newsreels.[64] The *Guntersville Advertiser & Democrat* encouraged residents to attend the screening because, even though Guntersville Dam was not yet under construction when the film was released, "in all probability a big dam will be under construction in this vicinity at an early date [and the] picture will be of unusual interest to our people at this time." Later, the Ritz featured a special *Time* magazine-produced newsreel, *March of Time*, about the TVA. Newspapers billed this particular newsreel as the "impartial and well-rounded-out picture story of the TVA, the great agency of human uplift in the southeastern states."[65] The *Limestone Democrat* supported the feature for its ability to illustrate the TVA's benefits.[66] Florence's Princess Theater showed a film in October 1935 called *Electricity on the Farm*, intended for farmers who may have been slow to adopt the new electricity. TVA representatives were scheduled to be present at the screening, and residents were encouraged to visit the model kitchen, staged in the courthouse basement.[67]

The TVA's Information Office staff was not the only ones making movies to advance the agency's cause, nor were affected communities in the rural South the only audiences viewing TVA films. Hollywood soon became involved, bringing bigger budgets and higher-profile directors, like Pare Lorentz. His 1938 film, *The River*, about catastrophic flooding on the Mississippi River and subsequent damage to homes, roads, and farms, was a joint effort— produced by the FSA and distributed by Paramount—and was reflective of the government's initiative to put filmmakers back to work after the Great Depression. One of the earliest collaborations between a Hollywood studio and the government, it was an unprecedented move that helped elevate the TVA's national profile. The film was important in its own right and is generally considered among film scholars to be a landmark documentary. Lorentz asked Charles Krutch to relay an important message to William Sturdevant: "We are going willy nilly to show *The River* in every city in America and if Paramount does the job they are supposed to do, 50,000,000 theater goers will know a little bit more about the TVA by the end of the next 6 months."[68] Sure enough, the movie was shown in roughly 4,200 theaters across the country. The film ended with a bit about how the TVA was doing work to prevent Mississippi River catastrophes from happening in the Tennessee Valley.

Radio and film were important aspects of the Information Office's communication plans. But the foundation of their mediated strategy was print media. It offered them the widest variety of platforms from which to tell multiple aspects of their stories. Informational pamphlets, magazine articles, and, most importantly, newspaper articles catered to different audiences, allowed staff writers to work with journalists to create stories relevant to their type of publication, and provided a forum for daily updates related to all things TVA. It was through print media that the TVA's messages became truly ubiquitous in the Tennessee Valley.

The Office of Information routinely disseminated informational pamphlets to those who wrote to them, and pamphlets were available to people who visited dam sites. During the years of construction at Pickwick, Wheeler, and Guntersville Dams, the Information Office distributed just over 1.6 million informational pamphlets to Tennessee Valley residents, visitors to dam sites, and others around the globe.[69] "The Development of the Tennessee Valley" was one such pamphlet. It contained pictures of Wilson and Wheeler Dams, Muscle Shoals and the surrounding areas, a map of existing and proposed TVA projects, and imagery of the farm life that had been transformed by the TVA's presence in north Alabama. Graphic Arts staff made frequent use of Charles Krutch's photography, which offered side-by-side photographs of gullied, eroded land and the same area after the TVA's intervention, which included teaching farmers about appropriate fertilizer and land use. Pamphlets showed farm wives cheerfully using an electric range for the first time as well as the implementation of community refrigerators.[70]

One pamphlet called "The New Deal's Challenge is America's Opportunity!" took the form of a "study guide." This document was essentially a compilation of frequently asked questions about the TVA and covered topics including the history of Muscle Shoals, a justification of the government's interest in rural power, the TVA's yardstick rates, reservoir clearance, and the TVA's construction projects. Here was the TVA's attempt to generate interest among teachers, students, speakers, and club groups. The TVA targeted women with this booklet through the use of human interest stories and an approachable writing style. When describing the booklet's role in overall public information efforts, Sturdevant said, "Every effort is made to INTEREST and INSTRUCT the reader, incidentally furnishing him with ample grounds for argument for continuing the TVA for national reasons."[71] Reading the booklet would help audiences understand the TVA's sweeping plans for the South, as the Tennessee Valley was now "where dreams come true."[72]

The Information Office estimated that 1,676 magazine articles, or an average of one per day, were published about the TVA from 1933 to mid-1938.[73] Information Office writers sent out press releases at almost the same speed. Press releases were to "avoid trivial material, be largely confined to matters originating from definite Board action, be wary of the future tense and all should be newsworthy."[74] From 1933 to 1938, the Information Office distributed 672 press releases to their extensive mailing list of 17,893 individuals, groups, and newspapers. That number included 530 members of Congress, 681 county agents, 2,178 home demonstration agents, 178 Washington correspondents, 8,802 libraries, 75 magazines, 47 foreign newspapers, 2,038 US newspapers, 2,135 Tennessee Valley newspapers, 979 news indexes, and 86 labor publications.[75] The TVA was not the only New Deal agency to generate so much media attention, nor were they the only New Deal agency to rely on strategic communication to publicize their efforts. It was important for the public to know about these new agencies; the more people who knew about the agencies, the more people who could presumably be served. The Civilian Conservation Corps, the Works Progress Administration, and the Farm Service Agency, among others, had staffed information divisions to promote their work.[76] But the Information Office of the TVA was particularly active.[77]

In addition to their extensive press releases, staff writers also accompanied magazine and newspaper writers on trips to visit the dams and communities in which they were built. Serving as local community guides, the

Information Office staff was able to offer context to the many journalists asking for access to the buzz surrounding the Tennessee Valley. Charles Krutch remembers that print media were "avid for stuff" about the TVA, and there was "no difficulty at all" in finding people who wanted to write about this massive experiment in regional development.[78] Most writers from out of town who came to visit TVA projects failed to bring along a photographer, so Krutch's pictures most often were the ones used to accompany their stories. He also ended up serving as a TVA guide of sorts, escorting not only journalists but newspaper editors and book publishers around various TVA sites to get a glimpse of what the agency was doing.[79]

Generally, magazine coverage highlighted change in the Tennessee Valley taking place because of TVA programs. *Fortune* magazine published "Work in the Valley," which opened by summarizing varying opinions about the TVA. They mentioned the difficult work of the "undertaker to whom has fallen the job of removing several hundreds of the coffins, at $20 a shot," but also mentioned a hypothetical Wheeler sharecropper who was upset because "TVA has bought the land he farms right out from under him. . . . What is he to do?"[80] The article went on to further articulate the problems faced by sharecroppers: "You can't practice a proper rotation of crops if your job, as a tenant, is to wring the very gizzard of land you don't own with cotton and cotton and more of it." Arguing that the poor socioeconomic conditions plaguing southerners was not fault of their own, the authors went on to tell readers that Tennessee Valley residents "stew in the juices of such disadvantages as their forbears and their fellows themselves."[81] Another example of magazine promotion comes from the *Harper's Magazine* article "Teaching Grandmother How to Spin." In explaining the impact the TVA had on the Tennessee Valley, it referenced the "competition" between the first two cities to receive TVA power—Athens, Alabama, and Tupelo, Mississippi—to see which one could use the most electric current.[82]

Though the TVA was a program of national significance, the Information Office did not ignore the importance of connecting with local audiences. National media messages such as those found in newsreels, magazines, and radio shows reached some of the rural southerners who lived near the Tennessee River. But newspapers—particularly local weekly newspapers—offered the TVA the opportunity to diversify their strategy and strengthen their public message. The TVA's relationship with newspapers, especially local newspapers, bears special attention because of the extent to which TVA officials worked directly with local newspaper editors and courted high-profile journalists, recruiting TVA loyalists to support their projects in print.

5

TVA Newspaper Narratives

At first, Barrett Shelton was suspicious of the TVA. A lifelong resident of Decatur, Alabama, and the editor of the *Decatur Daily*, the town newspaper started by his mother and father, Shelton desperately wanted to see economic improvements in his hometown. In the early 1930s, Decatur was still reeling from the 1920s railroad strike that resulted in the closing of the Louisville & Nashville railroad shops and a subsequent economic downturn. Like other small towns across the country, the Great Depression hit Decatur hard: seven of Decatur's eight banks closed after the stock market crashed. Shelton's people had been "ill-fed, ill-clothed, and ill-housed" for as long as he could remember. But his people were industrious, and he strongly believed in the ability of Decatur residents to turn things around for their city, despite their repeated failed attempts to do so. Shelton initially did not want the government to have any control over his life or his neighbors' lives, nor did he think a big government intervention would result in true change for the South. Like many others, he still believed the Henry Ford offer to invest in Wilson Dam was a missed opportunity for north Alabama. In fact, he had in his possession a letter from Mr. Ford, who had written his father about his intent to take over Wilson Dam. Nothing since then had even come close to helping solve their biggest economic and social problems.

Shelton's attitude remained unchanged even when the newly elected, but not yet inaugurated, FDR spoke to Decatur residents just a couple of stories below his office window. He saw with his own eyes the "acres" of Decatur residents who came out to listen as FDR promised to put the Tennessee Valley "on the US map." Did he really have a plan to help them? He did not say. But the desperate residents, living "in an economic situation where no man knew where in the world to turn; didn't know where his next meal was coming from," surely hoped he did.[1]

So, when David Lilienthal first came to visit with Barrett Shelton and four other Decatur community leaders, who wanted a better life but had a healthy skepticism of any "new" plan, the reception was not friendly. "Almost hostile" was the way Shelton described the group's attitude toward Lilienthal; they viewed him as an obvious outsider whom they doubted fully understood their plight and whom they assumed was there to impose his will on people who could not resist. Their first question was, what was he going to do for them? Lilienthal "leaned his chair back against the wall and a twinkle of a smile came into his eyes, and he said gently and firmly, 'I'm not going to do anything. You're going to do it.'"[2] Shelton said that comment made him pay attention and was the beginning of his eventual wholehearted support for TVA programs in north Alabama. It was support that would prove crucial in helping the TVA's public image there.

Lilienthal later wrote in his journals of his "grand experience" in Decatur.[3] After their first meeting, Shelton quickly changed his mind about the TVA and became one of its most ardent supporters, a personal attitude that spilled over into the editorial and news content of his newspaper. Thus began a professional relationship that sparked a sincere and loyal friendship. Before too long, Barrett Shelton and David Lilienthal became "two men on a mission" to change the way of life in the Tennessee Valley.[4] They had different avenues with which to effect that change. Lilienthal had the multimillion-dollar backing of the TVA, and Barrett Shelton had a trusting audience of readers who would be among the first to use power generated by the TVA. Lilienthal felt Shelton's role in supporting the TVA was crucial, saying he "was simply lyric. . . . He told me that 42 articles were being made in Decatur."[5] The *Decatur Daily* was one of many local pro-TVA newspapers. Thanks to it and the other small-town newspapers, the TVA enjoyed overwhelmingly positive public opinion in north Alabama, the very place the TVA needed to be accepted if its programs were to succeed. This strategy of leveraging local newspapers worked. Lilienthal said the economically disadvantaged Decatur residents made a smart decision, in part from paying attention to suggestions made in the newspaper: "Instead of crying about the rain which was coming because much of the farming land was flooded by Wheeler Dam, they used their heads. The results have been really amazing."[6]

Newspapers were an authoritative source of credible information. Despite low literacy rates in north Alabama relative to the rest of the country, weekly newspapers were common. Every county in the area impacted by the TVA had at least one. North Alabama was served by the *Florence Times*, the *Huntsville Times*, the *Decatur Daily*, the *Alabama Courier*, the *Limestone Democrat*, the *Albertville Herald*, the *Boaz Leader*, and the *Guntersville Advertiser & Democrat*. Subscriptions averaged roughly $1 per year. Given what's known about annual income and literacy rates among the rural population, it's unlikely that most farmers subscribed to the newspapers. But they were still affected by the messages contained in the pages.

Regional and national newspapers may have reached even fewer north Alabama residents most directly impacted by the TVA. But that does not mean that their articles were any less important to understanding the relationship between newspapers and the public's acceptance of the TVA. Newspapers were effective sources of information because they reached the opinion leaders in any given community, who in turn passed along important information to those who did not have access to newspapers. In 1944, political scientist Paul Lazarsfeld developed the "two-step flow theory" of news media influence, arguing that the news media's effects followed a two-step progression.[7] An individual read the news, then passed the information on to someone who did not buy or read the newspaper. The second person was influenced by the newspaper's coverage, despite the fact that he or she didn't actually read it. The newspaper's influence flowed from the reader to the nonreader. It was a groundbreaking study that revealed the extensive influence that news media carried.

The TVA suspected that 75 percent of farm laborers, 84 percent of sharecroppers, and 50 percent of owners read no current literature.[8] Even so, the newspapers probably influenced local opinions about the TVA because of media's direct and indirect effects. Following this theory, is quite likely that local newspapers served as an authoritative voice in the Tennessee Valley and helped shape public opinion about the TVA by framing issues in a way that was consistent with their own ideological slants. The TVA, in turn, provided the newspapers with plenty of fodder for their columns: news updates on construction and the frequent staging of newsworthy events. The coverage reminded the public how important the TVA was for their future success and assured them that the economic growth and standard of living enjoyed by Americans in other parts of the country could be theirs if they supported the TVA.

While talented engineers, construction workers, and architects working for the TVA handled the very real social, economic, and geographical changes taking place in the Tennessee Valley, newspapers handled another challenge: convincing people to trust the TVA without questioning the agency's purpose or goals or the program's far-reaching implications. During the 1930s, the TVA made headlines in every north Alabama newspaper, as well as in the many regional and national newspapers that reached a wider audience beyond the Tennessee River. Newspapers even reported on the TVA when there was nothing new to report; such was the case in a *Guntersville Advertiser & Democrat* story with the headline, "No New Dam News."[9] Politicians running for office supported the TVA in campaign speeches and full-page newspaper advertisements at both the local and regional levels.[10] Entire communities were publicly praised for embracing the cheap electricity the TVA supplied and for supporting the local economy by buying modern electric appliances such as hot water heaters, refrigerators, and washing machines. Editorials reminded those who disagreed with the TVA's presence in the Tennessee Valley that they were selfish and outnumbered.[11] It should be noted that the plight of the poorest Tennessee Valley residents—the share-croppers, tenant farmers, and squatters who were forced to relocate—was overlooked in newspaper coverage. Why did newspapers ignore those who stood the most to gain, as well as the most to potentially lose, in the TVA experiment? From more than 800 feature stories and editorials about the TVA placed in north Alabama newspapers in the construction era, only thirteen were devoted to family relocation, an issue that impacted over two thousand families in north Alabama. Though they tended to be more critical of the TVA, regional and national newspapers took a similar pro-TVA stance. Newspapers supported the TVA because of anticipated economic benefits for the area—and for the newspapers—and to assist the TVA in the best way they could, which was by shaping a positive public opinion. Minimizing, or omitting, coverage of the relocated helped with that goal. National and regional newspapers generally supported the TVA, too, though positive coverage was tempered with more critical analysis.

Two useful theories demonstrate how newspapers may have shaped public opinion about the TVA. One is the agenda setting theory, which asserts that the news media set the news agenda by telling news consumers what they should be thinking about.[12] In other words, by placing a story on the front page, the newspaper tells consumers that it is an important story and they should be interested and/or concerned about it. Conversely, ignoring a story communicates that it is unworthy of news media attention; it is un-important. In these ways, the news media set the agenda on societal issues. Newspapers held a great deal of power in shaping the public's understanding of what the TVA meant for the region. Local newspaper editors had the privilege of deciding what type of stories to print about the TVA: critical thought pieces that directed citizens to question the TVA's motives, or pieces that supported the TVA. Newspapers overwhelmingly favored stories about the TVA's benefits, especially the potential for economic growth, over stories asking readers to confront the harsh reality that faced many farmers and their families who were forced off their land to make way for the dams.

Framing theory also helps explain the influence of news coverage. Frames refer to the manner in which news information is presented to the public and how the media communicate specific perspectives on stories. The theory states that news stories are told in ways that fit the dominant values of the editor, publisher, and consumers. Therefore, framing both produces and reproduces specific perceptions and interpretations of a news event and, equally as important, news consumers develop their own understandings consistent with the manner in which the story was framed.

The combination of framing and agenda setting suggests the potential for news organizations to have powerful influence on how news events are understood by the public. The potential effects, then, are twofold. The public reads only about the events that newspapers deem important enough to warrant publication, and news stories are written from particular interpretive frameworks. It is unlikely that audiences will learn about stories that are ignored by the news media, particularly during an era in which newspapers were the primary news source. As a result, questioning whether and how north Alabama newspapers covered the TVA goes beyond simple quantification. The answers to the questions speak to the profound ability newspapers possessed to shape public response to an unparalleled social experiment that drastically changed the northern part of the state.

The TVA enjoyed a friendly relationship with local journalists as well, thanks in large part to the professional network and experience of William Sturdevant and Maurice Henle, the Information Office director and assistant director. Journalists worked closely with Information Office staff to craft articles about the TVA, at least part of the time specifically writing "to sell the TVA to the people of the Tennessee Valley."[13] For example, an Information Office staffer prepared an article specifically for the *Chattanooga News* that emphasized "the present trend of population, industry, and wealth toward the deep-water margins."[14] They also supplied "illustrations" with captions that supported the article.

The *Chattanooga News* was not the only outlet to rely heavily on Information Office staffers to write and photograph stories for publication. National newspapers also were courted by the Information Office. Journalist R. L. Duffus of the *New York Times* wrote favorably about the TVA while maintaining a friendly relationship with William Sturdevant. Duffus was a self-proclaimed TVA supporter, and as a writer for the *Times*, Sturdevant found him "especially worthy of attention" by the Information Office as "laudatory treatments of TVA" such as his were deemed "priceless assets to the public relations campaign."[15] Duffus once wrote to Sturdevant, "I think my paper is less enthusiastic about TVA than I am, but that makes no difference—it respects my private conscience and asks no opinions of me that I don't hold."[16] Before publishing one article about TVA programs in north Alabama, Duffus sent it directly to Sturdevant for review, saying, "Despite my objectivity I was moved by what I saw in the Valley to want to write an article that would rub your fur the right way."[17] Sturdevant's response was enthusiastic. He said, "You wrote of Norris Dam as a 'jewel.' I would like to apply the same description to your article on TVA."[18] He later wrote to thank him again, discussing the further impact his article to a national audience had on the TVA. After its publication, TVA dam sites found themselves "swamped with visitors" from all over the country. Sturdevant felt the increase in visitors was directly related to the article Duffus published, which idealized the TVA and its projects.[19] Duffus called Norris Dam "a dream . . . designed to change the life of millions," and he reminded readers that "a river as large as the Tennessee cannot be developed by private enterprise—the job is too big and there are too many uncertainties in it. Therefore, government must do the developing and private enterprise can step in where government leaves off."[20] Duffus was one of many TVA sympathizers who used their public platform to share information they learned in Information Office press releases and on tours of dam and reservoir sites. Though Duffus had ethical and professional responsibilities as a journalist to report objectively on government dealings, his own personal feelings, coincidentally or not, were consistent with the tone of his reporting about the TVA.[21]

Though local newspapers' support was crucial to north Alabama residents' acceptance of the TVA, sparking the interest of the rest of the country did not hurt. The number of visitors to TVA dam sites was included in each TVA annual report—a report on which Mr. Duffus coincidentally was a paid consultant in 1936, offering suggestions on the way such data was reported and which data were included.[22] This report was distributed not just to Congress but also to media organizations that requested it. Sturdevant thanked Duffus for his impressive work on the report, saying the TVA was "delighted" with his work, as the report "received a very good press, and even those newspapers which are vigorously opposed to TVA have seen fit to give it a fair review."[23]

Thanks to thousands of press releases, relationships cultivated with journalists and newspaper editors, and a staff prepared to respond to any negative coverage, newspaper stories about the TVA in local, regional, and national newspapers were overwhelmingly positive, serving to promote the TVA. Local newspapers minimized controversies, which were seen as threats to the TVA's overall aims, and national newspapers focused on big-picture issues and overlooked details that might have led to the public's understanding the full extent of the TVA's presence in north Alabama.

It comes as no surprise that north Alabama newspapers were supportive of the TVA. It was the first large-scale yet feasible attempt at revolutionizing and modernizing the area. In *The Story of TVA*, John Gunther writes that when all newspaper editors in the Tennessee Valley were asked about the TVA, "every reply was favorable, out of several hundred, except three."[24] This statement was confirmed by early TVA employee Fannon Beauchamp, who acknowledged that newspapers "supported us overwhelmingly."[25] North Alabama newspaper articles about the TVA generally revolved around specific themes. Newspapers wrote extensively about the TVA's economic impact, visits to TVA sites by prominent politicians, employment, rural electrification, dam construction, and the selfishness of resistance to the TVA.

Perhaps the most powerful set of articles in shaping public opinion addressed how the TVA would release the untapped potential of the South. The *New York Times* called it the "Ruhr of the South," and explained that the millions of dollars spent on the dams would "change the face of America."[26] The *Chattanooga Daily Times* said the South was "blessed by TVA's work."[27] In 1934, the *Decatur Daily* reported that the TVA's efforts would result in a "modern utopia" for its customers, whose lives would be transformed by electrification.[28] Moreover, thanks to the TVA, newspapers had a reason to look forward to the future: the TVA promised historic levels of economic development in the Tennessee Valley. Newspapers praised the towns and counties in which their readers lived with the message that theirs was a great region that would be greatly improved with some assistance. Articles romanticized the unruly river.[29] Those who did not see the potential for development were "foolish" as the TVA assured "a rapidly developing section of the South" that would make north Alabama "the greatest section in the South within a few years."[30]

Most north Alabama newspapers were effusive about the economic improvements soon coming to the region. The TVA was said to encourage "general, all-around, permanent development of this region" including supporting industries and rural electrification, and, thanks to TVA programs, the region would serve as a "shining example for the inspiration and education of the whole country."[31] Newspapers had high hopes for how the TVA would impact long-term development of the region and reported the TVA's promise of a "new social order to the mountains . . . social improvements for millions."[32] In March 1934, when the TVA was little more than a radical idea, newspapers hoped that the TVA would mean "a new expression of American life . . . if successful . . . [which] may sweep across the whole United States as rapidly as transmission lines can carry cheap electricity to every home and shop," and it also might "lift the people of the mountains and many in the valley from abject poverty and environmental frustration to a place of prosperity, happiness and usefulness."[33] The TVA meant a better life for those who had "always been poverty stricken and ignorant [as they would] soon be taking their part in the national life."[34] The TVA promised to "transform the mountain people from backward woodsmen to progressive citizens."[35]

Regional newspapers such as the *Chattanooga Daily Times* were also supportive of the TVA, though there were fewer articles about the agency in those larger, daily publications. News of the TVA made headlines in newspapers targeting smaller markets, too, including the *Atlanta Daily World*, the most prominent black newspaper in Atlanta. Articles about the TVA in the *ADW* focused on the TVA's policies regarding race relations, including the agency's promise of more equitable hiring practices for black workers.[36] The *Chattanooga Daily Times* closely followed the TVA's presence in Tennessee, particularly as it related to decreased electricity rates for Chattanooga residents and the potential for the EHFA to be headquartered there.[37] A 1936 article negated accusations of socialism, reminding readers that the TVA was not socialist because it was intent on "giving advice as to the best ways of doing things and leaving it to individuals and communities to decide for themselves how the advice works out."[38] The *New York Times* updated readers on the progress of construction at the close of 1933, claiming "harmony is in sight, hum of industry heard, money pouring in."[39]

Newspapers used the bandwagon technique, reminding readers that those who may have opposed the TVA were in the minority, going so far as to shame those who might have spoken out against the agency. Instead of focusing on what might happen if the experiment failed, or what was to be lost in the modernization of this area—land, taxes, homes, and an entire way of life—newspapers emphasized the possibilities for the area if TVA was successful. Articles, editorials, and advertisements reinforced the message the TVA desperately needed people to accept: TVA programs would be beneficial to everyone. How exactly it would benefit those who sacrificed to make room for the TVA, newspapers did not explain. Instead, they frequently reminded readers that whatever TVA projects were happening in their backyards was part of a much bigger plan that impacted more than just what they could see and more people than they could ever possibly know. For example, on a visit to Limestone County, TVA director Harcourt Morgan and Dr. L. N. Duncan of the Alabama Polytechnic Institute (later Auburn University) "urged Alabamians not to look at this or any other feature of the TVA development program from a local point of view but from their national aspects" as the agency was designed to benefit the entire country.[40] Chairman A. E. Morgan spoke of the "improved waterway traffic" not just for north Alabama residents but for everyone living along the Tennessee River.[41] Wheeler, Guntersville, and Pickwick Dams were but three pieces in a much larger puzzle, and newspapers communicated it would have been a disservice for anyone to have stood in opposition of what "everyone" felt was good for the Tennessee Valley.

One way newspapers generated public support with the bandwagon technique was by frequent mention of the other cities, towns, and rural areas that were supposedly benefiting from the TVA's programs.[42] Public attention directed on those cities was positive, and the praise showered on those areas accepting the TVA's aims was plentiful. Examples of other north Alabama, east Mississippi, and south Tennessee cities, such as Tupelo and Pulaski, that were enjoying rural electricity benefits from TVA programs reminded north Alabama residents that they were part of a program with a broader reach than their backyards. By December 1934, "21 counties in the state of Mississippi [had] organized to request TVA power," which "means that the towns of that state have been 'sold' by Tupelo's experience" and it was only a matter of time before Alabama cities would be asking for TVA power. Even the largest city to the north, Nashville, was interested in the TVA project.[43] The way newspapers wrote about the TVA, it sounded like everybody was in favor of TVA power, regardless of the costs. Even advertisements supported the TVA, sometimes when they had nothing to do with electricity or appliances. An August 1934 advertisement in the *Alabama Courier*, paid for by the local Dr. Pepper bottling company, simply stated: "Drink Dr. Pepper and cooperate with the TVA—we are."[44]

Newspapers made it sound like the public favored the TVA even before they had a chance to understand what it would mean to them. As early as July 1934, the *Albertville Herald* wrote about Roosevelt's plans to make electricity affordable for rural areas, a move the *Herald* translated as a policy that "has been accepted with overwhelming approval by both town and country everywhere electricity has been made available."[45] The *Albertville Herald* wrote about the "huge TVA project" coming to Guntersville, the dam, which was "a fine piece of news and," the newspaper assumed, would "be happily received by the people of this section."[46] Political leaders said FDR's work to create the TVA rural electrification program would "immortalize him and place his name in history as one of the great benefactors of mankind."[47] FDR was called "our best friend,"[48] and people were asked to support him in the 1936 election. The article said that so many people should support him, including "those who sold land to the TVA at TVA prices, and all merchants, automobile sales agencies, those who made profits from the sale of refrigerators and other TVA appliances, as well as the bankers whose deposits were swelled by the distribution of these funds, and those whose homes and farms have been saved by loans, and practically every person residing in this county."[49] Furthermore, political leaders who disagreed with the TVA, according to some, were in trouble. One Chattanooga representative was quoted as saying that "those in public life who opposed the Roosevelt power policies would be swept from office."[50] On President's Day 1936, the *Albertville Herald* called upon the Founding Fathers. They ran an editorial arguing that George Washington would have supported the TVA. According to the editorial, Washington "was progressive, and was always looking ahead. He demonstrated these characteristics by his accepting the leading role in the war of the revolution. . . . He would be looking to the future development of this natural resources of his nation."[51]

The newspapers also offered specific examples of people who saved money thanks to the TVA's rural electrification program. These stories personalized the TVA and bolstered the idea that the TVA offered better rates than their only would-be competitor, the Alabama Power Company, which lost a lengthy legal battle over the constitutionality of the TVA's power production and distribution goals. One newspaper ran a story about Athens resident Manion Mitchell, who owned "an electric refrigerator, range, water heater, radio, lights, vacuum cleaner, food mixer, toaster, percolator, waffle iron, electric iron, hair dryer, electric razor, wood-working machines, [and] heaters in [his] bedroom, kitchen and bath."[52] The article compared his power usage and rates with a family in Florence who used Alabama Power rates, saying "the difference between TVA power and Alabama Power Company is the difference between electricity working for a man and a man working for his electricity!"[53] This article was written in a way that made it difficult to argue with the assertion that the TVA provided a cost savings.

Newspapers continuously reminded readers that the TVA's presence was what the majority of people supposedly wanted. Municipalities held referendums on whether to keep private power supplied by Alabama Power Company or to allow their local governments to take over the distribution and selling of electricity generated by the TVA. Thanks to the promise of cheaper electricity and faster, more widespread access, people voted in favor of TVA power over Alabama Power.[54] However, just because a majority of voting residents agreed they wanted TVA electricity does not mean that all people were as enthusiastic about what that meant for other areas of their life, or that everyone had an equal opportunity to weigh in on the decision. Eventually, it was just such a community vote that caused the editor of one north Alabama newspaper that was initially critical of the TVA, the *Florence Herald*, to throw his support behind the agency. The editor formally and officially backed off his criticism in the May 22, 1936, issue in an editorial titled "Local Power Plans."[55] He still did not understand the "infatuation" Florence residents had with the TVA, and expressed his concerns that "the municipal ownership plan has generally been the badge of a hick town, and in the last 25 years a vast majority of such towns have sold their municipal plants to more efficient private companies. If our people think political operation will make them more prosperous and happy, let them go for it." In Chattanooga, politicians were warned they would be voted out of office for opposing the Roosevelt administration which, essentially, meant opposing the TVA.[56]

Construction efforts at Wheeler, Guntersville, and Pickwick Dams attracted the attention of local, state, national, and world leaders who, by their very presence in north Alabama, seemed to reinforce and support the newspapers' suggestions that the area had great development potential thanks to the TVA's efforts. The excitement of visits from important political leaders is obvious in stories about politicians and TVA leaders. Newspapers routinely noted when key visitors were in the area to supervise construction, check on the progress of the TVA, or explore what the TVA was doing as a model for development in other parts of the world. The TVA board of directors routinely stopped in north Alabama cities most impacted by the TVA's construction, reminding citizens of the TVA's importance and thanking residents for warmly embracing the program.[57]

Dignitaries from all levels of government were welcomed as readily as TVA officials. A few months after the TVA Act was initially passed, President Roosevelt visited Alabama. The *Albertville Herald* reported that "thousands" turned out to "greet the president and get a close-up of his now famous smile. . . . Probably never in the history of the state have such large and enthusiastic crowds turned out to greet a public official."[58] The newspaper went on to say that "everywhere he went the President was made to feel that the people of the country are behind him and his New Deal." In nearby Athens, newspapers promoted the President's visit the week previous to his arrival, and though residents were disappointed his train did not stop in Athens, "5,000 people lined the railroad tracks [and] had the pleasure of enjoying his magnetic smile and the graciousness of Mrs. Roosevelt as they stood on the rear platform of their car all the way through Athens and waved to admiring throngs."[59] Many residents were proud to have the TVA, and politicians and governmental leaders were proud that the people of the South so readily embraced the agency. At a time when the only other major story coming from Alabama focused on the Scottsboro Boys—a case in which nine black boys were erroneously accused of rape—the attention of political and government leaders shone a more positive light on the region. Reading pro-TVA newspaper stories further legitimized the TVA and gave the impression that *everyone* was in favor of the progressive agency.

More important than political visits or the promise of long-term development was the most immediate and practical benefit of the TVA: jobs for unemployed people or those barely surviving on small government stipends. Newspapers frequently mentioned the number of people who were employed by the TVA, which was another indication of how the news shaped the TVA agenda and framed its interpretation. It is hard to argue with any entity promising hundreds of jobs for people of the Tennessee Valley, and the presence of dams would mean local employees would be needed to work short- and long-term jobs. Newspaper announcements encouraged those interested in working for the TVA to apply.[60] People were put to work in mostly unskilled labor positions, such as clearing land that was soon to be flooded. This "herculean task . . . (would) require about two years to complete. All trees, undergrowth and debris of every nature is removed from the lands to be inundated."[61] The increase in employment meant an increase in just about every other economic aspect of the valley, too, including the ability to purchase electricity. Dam construction meant people were moving to the Tennessee Valley for jobs, which was previously unheard of.[62] The arrival of engineers and other skilled workers in the valley was newsworthy, though the *Chattanooga Daily Times* once published a report on a speech by Vanderbilt University professor Donald Davidson, a known figure in the agrarian movement, who suggested that the TVA should focus more on hiring workers from the South as opposed to outsourcing skilled and intellectual talent from other parts of the country.[63] This critical voice was a minority among newspaper coverage of new jobs provided by the TVA. According to the *New York Times*, the TVA's "main purpose . . . [was] to build up the permanent social and economic prosperity of the Tennessee Valley," a feat that would be "accomplished by using the people of the region on the job."[64] The newspaper later claimed that four million would be given jobs instead of relief.[65]

Devoting a considerable amount of media attention to the issue of rural electricity was intentional. Both the Information Office and Lilienthal worked to make sure the issue stayed salient through press releases, speeches, and other mediated coverage. The TVA needed to sell electricity to rural Alabamians in order to be truly successful, and one communication method the agency chose to meet that goal was to frequently and consistently remind citizens just how good they had it with TVA power. For example, one editorial wrote of the benefits of electricity for even the rural farmers who likely had no means to afford it: "Electricity on the farm is not a luxury but an economy. . . . Electricity makes your labor more productive: the one-mule farm becomes an electric farm."[66] Residents were encouraged to quickly start using TVA power, as soon as the lines were constructed and stretched out to the most far-reaching corners of the rural South.

Among the TVA's many promises, cheap electricity seemed to generate the most excitement. In addition to making farm life easier, business leaders expected to see an increase in industry.[67] "Attractive rates" on electricity, they believed, would mean benefits for business.[68] In fact, rural electrification was discussed in the media as if it was just as important, if not more important, than the promise of jobs and increased industrial development in the South. Although Alabama Power Company had provided electricity to the area for some time before the TVA's presence, the TVA promised more extensive electrification at rates that were a fraction of the cost of electricity supplied by Alabama Power. A 1934 *Alabama Courier* article compared rates in nearby Tupelo, Mississippi, "before and after TVA." A local homeowner whose previous bill from a Mississippi power company was $2.30, but his first bill for TVA power was 75¢. Another homeowner's bill dropped from $11.26 to $4.77. Business owners who relied on electricity saw a drop in rates, too. One department store reported a significant drop in its electricity bill, from $65.14 to $23.69 in the course of one month.[69] Clearly, access to TVA power meant substantial savings for those who used electricity. It was hard to deny the immediate economic impact of the program, especially given that money saved on electricity could be spent on something else, thereby improving local economies.

Part of the appeal of the TVA was its promise to provide electricity at an affordable cost to those who had never had it. Most citizens in areas directly affected by the TVA did not have electricity at the time dams and reservoir areas in north Alabama were constructed, despite the fact that Alabama Power and other private power companies were well established in the state. As late as August 1936, in fact, Alabama ranked thirty-fourth in the percentage of farms that had electricity.[70] One *Guntersville Advertiser & Democrat* article asserted that the cheap hydroelectric power produced by the new dams would be helpful not only for farmers, but would also cure the unemployment problem and would bring about a "rebalancing of the national populations between cities and rural sections."[71] The newspaper also reminded readers that the "consumption of electricity has long been recognized as one of the best business barometers . . . when industry really does come back, the power curve will be the first to show it."[72] As the TVA built more lines extending to more remote areas of the counties it served, the newspapers reported the progress. Newspapers touted the many benefits of cheap electricity to its readers, heaping praise on the TVA in the process. This type of coverage is an example of the agenda setting function of mass media. Newspapers directed readers' attention to the low electricity rates promised by the TVA, encouraging them to consider the benefits of TVA programs rather than the sacrifices made to facilitate cheap electricity.

Rural electricity really meant the modernization of the South; with access to affordable electricity, the South would literally be brought out of the Dark Ages. Newspapers acted as cheerleaders for valley residents, encouraging them to use power and praising them when they adopted it. Limestone County created a special "club" for families who used over 400 kilowatt hours per month, publicly displaying the families' names and discussing them in the newspapers.[73] It is possible that the desire to see their names in the newspaper spurred at least some residents to get electricity. By 1935, the *Limestone Democrat* was praising Athens in a full-page ad for "leading the way" with a "year of progress with TVA power."[74] Residents were strongly encouraged to use electricity, because the more electricity residents used, the cheaper the rates were for everyone.

The TVA years were hailed as revolutionary. A full-page photo spread in a November 1936 issue of the *New York Times* showed how rural residents across the South were using their new, cheap power. From beauty shops using curling irons to farm homes using electric ranges, water heaters, and refrigerators, TVA electricity was framed as an innovation that brought the rural South into a "new era."[75] Farmers were quoted in the *New York Times* as saying, "I don't see how we did without (electricity) so long. . . . The refrigerator and the washing machine make it a lot easier on the women folk, too."[76] Wilson Dam area farmer Owen Whitlock, one of a handful of farmers who had gone from sharecropping to owning land, was quoted on the impact of having electricity on the farm as "just like stepping from sorrow to sunshine."[77]

Folks accustomed to hauling ice blocks for refrigeration and gathering wood to maintain stove fires had to learn how to use electricity. In an effort to demonstrate the benefits of modern appliances, home extension agents held demonstrations at TVA showrooms, which were located inside businesses that sold appliances. This primarily encouraged the sale of kitchen-related appliances and educated homemakers about the promises of modernization, spurring them to accept help from sources they'd done without for a very long time. Appliance sales were crucial to the TVA project. Not only were they useful in getting people to use more electricity, but the more appliances businesses sold, the better it was for businesses and newspapers, who profited from the extensive advertising. Newspapers encouraged residents to attend these demonstrations.

The presence of electricity boosted the local economy through the increased sale of electric appliances. Advertisements in newspapers reflected the increase in interest in appliances, particularly in Limestone County. Appliance retailer Sam M. Bowen Company's bandwagon technique appeared in an advertisement in the *Limestone Democrat*, run four times in a 2-month period, boasting that it "looks like *everybody's* [emphasis theirs] buying a General Electric Refrigerator!"[78] The ads were not far from the truth. The *Huntsville Times* reported that in 1933, before the TVA's arrival, 9,785 refrigerators had been sold in the entire state of Alabama (compare that to the more than 200,000 refrigerators were sold in New York during the same time period).[79] But, after the TVA started providing cheap electricity, appliance sales started to climb. In August 1936 in Florence, Alabama, local businesses sold 132 appliances totaling $17,932; even in the rural areas of the county, appliance sales were steady.[80] During the same month in towns in the nearby Wheeler reservoir area, a total of 265 appliances totaling $32,928.36 were sold.[81] By April 1935, Athens had distinguished itself as a leader among cities using TVA in terms of the number of appliances its residents were buying, setting all-time sales records for the city and returning impressive sales numbers compared with other TVA cities.[82]

It is clear that Athens residents readily adopted the concept of modern, electric kitchens, and they did so at least partly in response newspapers supporting use of electricity purchase of appliances. Sales companies made it as easy as they could for families in north Alabama to buy appliances. At the time, they were still relatively expensive, especially for poorer farming families. However, as stories and advertisements reminded readers, financing was often available.[83] U. G. White Hardware in Athens sold Westinghouse refrigerators for $79.50.[84] The *Huntsville Times* published advertisements for a Faultless electric washing machine costing $44.50 in 1934.[85] Sam M. Bowen Company sold General Electric ranges that promised a "break from hot weather cooking" for $149.50.[86] Considering the low per capita income of north Alabama residents during the 1930s, these prices were steep, making appliances initially unaffordable for many living on farms. To assist residents with purchasing appliances, the EHFA was available to help finance the purchases.[87] Reminders of government assistance combined with the frequent advertisements for affordable appliances framed the TVA as a government agency that would modernize farm homes across the South.

An increase in appliances sold also meant more use for electricians, another way to put skilled laborers back to work. The *Alabama Courier* ran a long series of advertisements each week for several months from the local Athens Electric Service Shop, promoting their appliance repair services. Their February 20, 1936, advertisement summed up the attitude of many: "TVA is Our Papa—Let's All Help Papa."[88] Everyone seemingly wanted to do their part to welcome the TVA and the cheap electricity it offered.

Advertisements for appliances often ran next to news articles about appliance retailers. In this way, there was a strong connection between advertisers and positive press about the TVA. For example, Henry Neal Junior's appliance store in Athens placed an advertisement in the *Limestone Dem-*

ocrat on September 24, 1936, which was coincidentally the same day the newspaper ran an article about his relocation to a new venue in Athens.[89] Articles informing residents about increased accessibility of electricity ran alongside advertisements illustrating the wonderful benefits of electricity and a modern home. Local businesses and industry benefited from the increased sale of electricity, and newspapers benefited from the advertising dollars spent on promoting new products.

Advertisements were not just selling appliances and electricians' skills. Ads also sold people specifically on using electricity provided by the TVA. In the months leading up to the TVA takeover of Athens's municipal power distribution system, a series of ads from the Edward E. Word Electrical shop told readers to "Be Ready—Prepare for the TVA."[90] These small but frequent reminders of the TVA's presence encouraged people to embrace the new, cheap electricity once it was available. "Reddy Kilowatt," a stick figure with a round face designed by Alabama Power to symbolize electricity, was eventually co-opted for TVA promotions. An *Alabama Courier* ad from 1934, promoting "TVA: Electricity for All," described how helpful Reddy Kilowatt was to modern families. The ad touted: "Let Reddy Kilowatt do it for you—he'll cook your meals, freeze your ice and do most anything 'cept set the table and play bridge."[91] Another series of ads suggested Reddy Kilowatt "would like to do those tasks you have to do. All Reddy Hot, Mr. Kilowatt."[92] Advertisements like these worked together with advertisements for appliances and electricians to encourage the public to use electricity daily. Reddy Kilowatt made electricity seem approachable, even fun, which must have alleviated apprehension among those who may have been unsure about adopting electricity in their homes. These advertisements helped newspapers position the TVA as an agency designed to help all families through the process of modernizing.

Electricity would be good for everyone, but not everyone had the privilege of using electricity without sacrifice. For farmers who had little to no disposable income and who were barely keeping their families fed and clothed, purchasing appliances was out of the question. One example of a resident who may have heard about the benefits of electricity but would have been unable to afford it, or the appliances to take advantage of it, was Wheeler reservoir resident Henry Lucas, whose house was so cold, he froze to death one night. His children, who survived, went to live with neighbors.[93]

The TVA's electricity did more than just modernize the South. Increased availability and affordability of electricity jump-started the economy through appliance sales, appliance repair, and increased advertising revenue for newspapers that ran ads from local electricians and appliance retailers. Another side effect of TVA electricity was an increase in profits for municipalities who moved away from relationships with private power companies, instead voting to manage their own utilities. This controversial decision was at the heart of the heated legal debate about the constitutionality of the TVA. TVA sold electricity at wholesale costs to municipalities and cooperatives, which in turn would sell it to residents for a profit. This resulted in serious cash gains for cities; for example, Athens, the first city in Alabama to use TVA power, earned $6,500 from selling TVA power in 1 year. The *Limestone Democrat* gave credit to the mayor, R. H. Richardson Jr. and the residents who supported the TVA by buying appliances that increased electricity sales. The small town of Athens was hailed as "one of the most progressive cities in north Alabama, [which] never failed to show a profit."[94]

The *Democrat* also praised nearby Tupelo, Mississippi, demonstrating that Athens was not the only city adopting TVA power.[95] According to newspaper coverage, nearby cities like Tupelo scrambled to sign up for TVA power as well.[96] Newspapers made it seem like neighbors in other areas were enthusiastically and unquestioningly welcoming TVA power.[97] And, with profits like those in Athens and Tupelo, it's no surprise that everyone was on the TVA bandwagon. At least that's how the story was framed.

It would not have been feasible to communicate every detail of every aspect of the TVA to the residents in the Tennessee Valley. The Information Office primarily focused on sharing that which shed favorable light on the TVA, including updates on construction and economic impact. Two potentially contentious issues were noticeably absent from press releases: the removal of families and displacement of graves from the reservoir areas. Though there was a good deal of publicity about employment opportunities for men, TVA press releases never mentioned the thousands of people in north Alabama who were living on land purchased by TVA, nor did the TVA publicize their careful and respectful work on grave removal and interment. The TVA also did not place stories about the steps it took facilitate the migration of thousands of families from the riverbanks to outlying areas or the fact that they often wound up with less land and poorer soil quality.

One would think that with all the newspaper coverage of the TVA, notices about removal and relocation would be prominent and frequent. However, only one relocation-oriented article appeared in Wheeler area newspapers, buried on page 4 of the September 26, 1935, *Limestone Democrat*. The short hundred-word article asked "property owners in the Wheeler Dam reservoir to move all buildings below 556 contour off the property by October 31, 1935." Where the residents were supposed to go, or how they were supposed to get there, was not mentioned. Interestingly, this article was positioned directly underneath an article praising the citizens of Athens for leading TVA cities in electricity use. "Being a leader in all good things is one of Athens' outstanding characteristics and is not at all surprising to find her 'leading the league' in residential use of electricity."[98] Newspapers did not acknowledge the hardships families faced in relocating. Instead, the issue was given minimal attention, couched among stories praising the TVA and residents for so readily adopting TVA power.

More coverage of relocation was afforded to residents of the Guntersville reservoir area. The *Guntersville Advertiser & Democrat* and *Albertville Herald* ran similar stories on March 11 and 12, 1936, encouraging residents to attend a meeting about relocation that would be sponsored by the Alabama Extension Service. This meeting would help residents understand their options for relocation, acknowledging that the construction of Guntersville Dam "will cover quite a bit of our good farm land and will disturb both the social and economic conditions of the farm families affected."[99] The articles both expressed hope that the farmers, who were "assets to Marshall County,"[100] would remain in the county postrelocation. Both the *Guntersville Advertiser & Democrat* and the *Albertville Herald* published more articles announcing the assistance farmers would have from extension agents in finding new places to live.[101] However, no articles announced any kind of assistance or relief for tenant farmers or sharecroppers, the classes of people who did not own their land and received no financial compensation for having to move.

Despite newspapers telling citizens to trust the TVA, there was reason for residents most directly impacted by land purchase efforts to be skeptical. There is evidence that farmers may have been taken advantage of in selling their land. The *Alabama Courier, Limestone Democrat*, the *Florence Herald*, and the *Guntersville Advertiser & Democrat* all ran advertisements placed by the Tennessee Valley Landowners Mutual Aid Agency (TVLMAA) in December 1934 and January 1935 that detailed the TVA's land purchasing efforts in the Wheeler Dam area and again in 1936 for Guntersville Dam.[102] TVLMAA was created by J. W. Knight to benefit landowners, particularly by collecting information for them and helping them sell their lands at fair market value.[103] Rather than trusting the TVA to treat landowners fairly, the TVLMAA helped landowners negotiate with the TVA. Despite its obvious benefits to landowners, the TVLMAA received no coverage by newspapers; the TVLMAA instead placed their critiques of the TVA's operations related to land purchases as advertisements. Internal documents revealed that the agency believed the TVA was "purchasing lands at less price than they had expected . . . [and] some owners have been offered as low as $7/ acre in the Wheeler Dam area," and the TVLMAA worked to remind citizens that "just compensation means the full fair value of the land, not TVA land agents' values."[104] Such outwardly critical advertisements like these were not common, and no news articles were written about the TVLMAA's ads.

Local newspapers, then, framed the TVA's land purchase efforts as positive, or at least not harmful, to north Alabama residents. By arguing that the TVA hurt no one and that some would be helped, through land acquisition, newspapers further contributed to the dominant public opinion that the TVA was helping the Tennessee Valley meet both short- and long-term development milestones. The relocation of 2,500 families from the land they purchased, while controversial, generated surprisingly little media attention.

The *New York Times* offered a more balanced view of land purchasing and family removal, though their reports were hardly critical. One of the earliest articles about the TVA, from July 1933, indicated that the agency would "allow the small home-owners of [the] community plenty of time to acquire other homesteads."[105] This statement proved to be somewhat accurate; despite giving residents a deadline to vacate the riverbanks, the TVA did frequently offer extensions of that deadline to families who were having difficulty finding new places to live. Later, the *Times* acknowledged the need for many residents to adjust to a new way of life after relocation, which would "be made, it was indicated, by the encouragement of local industries, especially for home consumption, and resettlement of families displaced by the establishment of national forests, the construction of reservoirs or by other causes."[106] The readjustment of families after relocation proved to be more difficult, as will be discussed in a later chapter.

Inevitably, a handful of media messages contained critical commentary about the TVA. When this happened, the Information Office responded, generally to counterargue their points. David Lilienthal suggested to William Sturdevant that his office should be in the habit of responding whenever newspapers made what he claimed were "inaccurate or misleading statements." As a result, staffers read every article they could find about the TVA for errors and inconsistencies in coverage. Though he did not expect newspapers to follow up on the corrections or retractions Information Office staff suggested, Lilienthal felt that "the cumulative effect of repeatedly correcting them would be very good in tending toward more careful scrutiny."[107] After understanding what the relocated residents in north Alabama experienced, one may argue that those publications may simply have been offering an alternative point of view in a sea of dominant pro-TVA coverage.

But the Information Office found themselves writing to numerous publications from the *Birmingham News* to *The Atlantic* to correct information they deemed was printed in error.[108] Occasionally, those "errors" were reported to Information Office staff by loyal newspaper editors across the country. *Alabama Journal* editor C. M. Stanley mailed Lilienthal a copy of an editorial from the January 29, 1936 edition of the *Mobile Register*. Stanley wrote, "it occurred to me you would be interested in seeing the lengths to which these people will go in their treatment of an Alabama asset."[109] The editorial had criticized the TVA for supposedly failing to share information about salaries. W. L. Sturdevant replied to the editor at the *Mobile Register*, reminding him that all salaries above $1500 per year were included in the TVA annual report. Sturdevant went on to quote an article from the February 1 issue of *Business Week* called "No TVA Secrecy," which praised the TVA for its transparency. Sturdevant closed his memo with a request that, "in fairness to your readers ... [give] this letter the same editorial prominence you gave your original unfair charge."[110]

Sometimes, the Information Office could not refute criticism printed in the newspapers. Early on in the TVA years, the *New York Times* suggested that the TVA was taking too long to put people back to work, as they promised they would. Writer W. G. Foster claimed that "apparently [the TVA] is not thinking of the hungry people of the present moment" in claiming it would be another 6 months before Tennessee Valley residents would see an influx in dam and reservoir-related land-clearing jobs.[111] Another article wondered whether the TVA's "vision of social planning," which they reminded readers was a "great experiment," would succeed.[112] Commenting on the hardships of the relocated and the disenfranchisement of black residents, *Fortune* magazine noted that for tenants in the Wheeler reservoir area, "your landlord gets the money [and] all you get is moving orders."[113] Though they admitted the TVA was doing what it could for relocated residents, "the share cropper and the Negro are two profoundly painful problems that TVA . . . has no constitutional power to solve."[114]

But the glowing news of TVA influence rarely dimmed in the smaller local newspapers. Only nine articles or editorials across all north Alabama newspapers during the mid-1930s directly criticized the TVA, and even they were tempered with positive comments about the agency. A *Huntsville Times* article reported that Alabama state Representative Archibald Hill Carmichael had bemoaned the fact that TVA headquarters would be located in Knoxville as opposed to the Florence area as originally promised, which would have brought more industry to north Alabama. The article closed by saying the representative "was in full accord with the broad program of the TVA."[115] An *Albertville Herald* editorial warned readers of the amount of money Congress would be expected to pay for the dams and cautioned the TVA in dealing with the "proud and touchy" people of the Tennessee Valley."[116] Even so, the article closed with praise for the TVA board of directors. Criticism of the TVA was generally mild and when expressed was generally countered with something positive.

The *Florence Herald* was an outlier to the others' constant praise of the TVA, and most of the criticism about the TVA came from this newspaper. The editor once wrote "there has never been any secret about this newspaper's determined and conscientious opposition to government operation of business enterprises,"[117] suggesting not only opposition to the TVA but also to governmental influence over business. In August 1935, when land-clearing efforts were slowing down, the *Florence Herald* brought up the issue of many local residents left without income thanks to their land clearance contracts ending, apparently wishing to "give readers something to think about."[118] Furthermore, the *Florence Herald* was the only newspaper in the Tennessee Valley that supported the Alabama Power Company in its legal battle with the TVA. Here is a unique example of the power of agenda setting in which the *Herald* was an anomaly among similar newspapers that praised the TVA. The newspaper routinely criticized the TVA in an effort to provide its readers an alternative viewpoint. Out of all north Alabama newspapers active during the 1930s, the *Florence Herald* was the only one that routinely offered criticism of the TVA.

Newspaper coverage of TVA glowed like the light bulbs everyone would have in their homes once electricity was widely adopted. Editors and publishers took every opportunity to extol the seemingly countless advantages the TVA brought to north Alabama. Modern homes, healthy economies, cheap electricity, high employment, and increased civic revenues were said to all result from one agency: the TVA. There was little criticism of the TVA in local newspapers in the early years of the agency's existence, just as there was little attention focused on any controversial aspects of the TVA program.

Local newspapers largely ignored the realities of life for families living on the riverbanks who sacrificed much in the name of progress. They generally did not criticize the TVA, nor did they bring to the public's attention any of the hardships that were imposed upon the relocated, despite their severity. However, these areas of the TVA project arguably were both newsworthy and worthy of critical attention. The TVA's land purchase efforts created a land shortage, overcrowded housing, and land conditions. Family removal did little to improve race relations and further oppressed a marginalized population who had very little ability to counter the government's plans to develop the region. These immediate consequences were not raised in mediated discussions of the TVA, which further contributed to shaping a positive public opinion about the agency. The TVA's stated ultimate goal was to promote the "well-being and security of the people of the area,"[119] but did they fulfill that goal as quickly as newspapers would have readers believe?

While newspapers reported regular updates on land purchase issues, they rarely reported on the people directly affected, and in some cases harmed, by those purchases: tenant farmers who depended on landowners for cropland and owners who were losing valuable acreage due to the land shortage that TVA construction created. Reporting only on the assistance farmers received from the TVA is yet another way newspapers shaped public opinion about the TVA. By omitting information about the plight of those who did not own land but still had to move, newspapers made it sound like everyone who needed to relocate would be equally assisted. This was simply not the case. Because of the resulting land shortage created, in many cases, farmers had to downsize their operations, which meant tenant farmers and sharecroppers were left with nowhere to farm. Most people who had to relocate did not own their land and were not eligible to work with extension agents to find new farms; they lacked money to buy their own land before the TVA, nor would they have such an income until much later. By paying little attention to the problems TVA caused landowners along the river and, more severely, sharecroppers and squatters, north Alabama newspapers focused their readers' attention on the TVA's promises and possibilities rather than its potential problems. The public agenda was set in favor of the TVA, even if there were skeptics did not agree that the TVA's presence was an entirely positive thing.

6
Minimizing Threats and Criticism

Even though the TVA Information Office had the ability to heavily influence the media's largely pro-TVA narrative, who could blame editors and journalists for supporting the cause with such positive coverage? Few, especially in north Alabama, would argue that the many housing, health, and social problems facing rural southerners needed solving. But the TVA was presented as the only solution. Newspapers did not engage in critical debate or balanced coverage about the issue of family removal and relocation, overlooking a newsworthy event in the process. Local newspapers essentially failed their watchdog function as an objective source of information. They were influenced by the TVA Information Office and personal relationships with TVA officials. Had local newspapers "followed the story" and talked with residents affected by family removal, they likely would have found some stories that contradicted the typical pro-TVA stance.

Positive coverage was important to the TVA mission. As one author noted, the TVA "could have the noblest dreams for 'the economic and social well-being of the people living in said river basin,' but the dreams would come to nothing unless the people were themselves persuaded that what was being done and proposed was good."[1] Newspapers constructed the narrative that the TVA was progressing through its plans with very few problems. The issue of family relocation, however, was more complicated: Problems experienced by the relocated posed a legitimate threat to the overall success of the TVA project. Instead of highlighting examples of families who bettered themselves through family removal, most newspapers opted largely to ignore the issue of relocation altogether. There were struggles and difficulties unique to those who were most crucial in making way for the TVA and evidence that their lives did not improve significantly or quickly.

This was not the government's only New Deal era attempt at population resettlement. Three other agencies—the FSA, the Civilian Conservation

Corps, and the Federal Subsistence Homesteads Corporation—attempt-ed to assist with resettlement as part of larger relief efforts.[2] However, the TVA's family removal and relocation program differed in that residents were forced from their homes for the creation of reservoir areas. Though most understood, and even on some level agreed with, their role in aiding the TVA's goals, residents still felt strong attachments to the land on which they had lived for generations and, in some cases, resisted moving.

Relocation from the reservoir areas was emotionally difficult for many residents. Some mourned the loss of the land like they would a death in the family. While renters did not own their land and were accustomed to relo-cating frequently, they rarely moved far because they, too, were emotionally attached to the area. John Bennett of Brown's Valley in Marshall County was one of the few in the Guntersville area who moved out of state. He purchased a farm in Viola Valley near McMinnville, Tennessee, but he did not seem to adjust well to his new home. Relocation caseworkers noted that he seemed to have "an attraction, home ties, or something in Alabama that he cannot sever."[3] Despite having a "good attitude" toward the TVA, many families found it difficult to move away from relatives who "they regret[ted] the necessity of leaving."[4] Readjustment to a new way of life was particularly difficult for families separated specifically because of the TVA's programs; this disconnection left emotional scars that remain today.

Luther Tidwell's family had difficulty relocating from their Guntersville farm and away from their friends who lived nearby. He said neither his mother, who was 55, nor his father, who was 60, were the same after the move. They both "grieved over leaving home."[5] Despite receiving $10,000 for their farm, "life at home was never the same, cause my mother was . . . happy go lucky and jolly . . . but after, she [got sick and] couldn't do any-thing. [She] never was able to hardly do anything but walk around after we moved. And she didn't live but a little while after we [moved]." Depression was common. It was difficult for some families to leave the only homes they'd ever known.[6] Similarly, Hazel Thompson's close-knit extended fam-ily, which included aunts, uncles, and grandparents, was separated by TVA family removal. Her father and mother had no choice but to move their family to Hazel Green, Alabama, about 50 miles from their previous home in Guntersville. Before relocating, Hazel had spent considerable time with her grandmother and cousins. But after the move, seeing them at all proved difficult because of the distance and the lack of a car to make the journey. She "missed 'em so bad, my sister . . . cried because we left! And, it was bad, you know, not to, to get away from your grandparents and not ever be close to 'em anymore."[7] Hazel remembered that her family had lived in the same area for generations and they used to gather under an old oak tree for reunions. After the TVA purchased land from her family, the oak tree and a nearby cemetery in which many of her relatives were buried were inaccessible to them.[8]

Some people's feelings were quite simply hurt over having to relocate. Gus Ross and his wife were both "in tears when they talked about leaving their present location." They had lived in their home since 1919 and were not happy about leaving after 19 years there. The caseworker said he "made multiple visits to 'correct the attitude' of Mr. R., (who) felt he was being pushed around because of his age and race."[9] Jeff Walker's wife was "at times . . . so overcome by her feeling of regret [of having to move] that she was unable to talk. [She was] deeply concerned over the necessity of giving up her home to which she has sentimental attachments."[10] And in some cases, residents' negative attitudes toward relocation were not communicated to the TVA caseworker. This was the case for Hazel Moore Thompson and her family. When talking about how much positive buzz surrounded the TVA, giving the perception that people always willingly left their homes, she responded, "Well, we minded! We didn't like it. I don't believe nobody really wanted to move."[11] Even though relocated families understood the promises the TVA made for substantial improvements to their lives, they were, on some level, upset about being told to find another place to live.

The poverty, housing conditions, health issues, and other social problems that were made worse by relocation resulted in despair among many who were forced to move. Caseworkers noted an attitude of "discouragement and indifference" for some families, which complicated relocation.[12] For example, the family of George Gilliam had economic issues and were "discouraged, [which] makes the situation serious."[13] Charlie Hilburn's family was "considerably below average with little energy, initiative or managerial ability and should be given careful attention. . . . Appears [they are] broken down morally and emotionally as well as economically. Unless family drifts back to mountain, will be difficult to relocate them."[14] With little incentive to move besides the fact that TVA caseworkers were constantly pressuring them to do so, those who were discouraged or desperate were in some ways a bigger threat to construction efforts than those who were extremely vocal about their displeasure. Feelings of helplessness brought on by forced removal contributed to further feelings of ill will.

Some residents penned heartfelt, desperate letters seeking assistance. These letters were presumably handled by the TVA Information Office, who worked to ensure that these negative sentiments were not widely publicized. C. M. Grant, a widow, wrote a letter in her own handwriting asking the TVA to give her permission to stay on her land.[15] Similarly, Mrs. Charlie Cloud sent a telegram to the "Lady in Charge = Moving People, TVA Headquarters," requesting, "dont [*sic*] come after me . . . place rented just below where we were."[16] Some residents bypassed the TVA completely and wrote directly to President Roosevelt or other government officials. Pat Toney wrote his letter to FDR, Senator John Rankin, and Senator George Norris.[17] In the letter, riddled with grammatical errors (Mr. Toney had never attended school), he asked for help. Mr. Toney's case was interesting, as he was the trustee for a 26-acre estate that the TVA had purchased for $3,507.50. He did not receive the money, however. It was applied to two previous mortgages, with the balance of $1,591 divided among Mr. Toney and three other parties. Whether or not Mr. Toney received a fair amount of money out of this deal is unclear. His letter dated January 7, 1936, said he'd "been trying for six mont [*sic*] to find some place that I could Buy I found places But not enuf [*sic*] money to come in." He was "asking . . . for mercy [as he was a] poor ol colored man." Eventually, Mr. Toney ended up renting land with the help of the Resettlement Administration. And Virgil Stone wrote a heart-wrenching letter to FDR on October 17, 1938, in which he said, "We are on the T. V. A. Land and have got to go soon and no place to go. No money to go with. We have never had no trouble of finding a man to back me up till the T. V. A. Bought the land on the Tennessee River and now I can't find no man that will carry me over for my debts. . . . I could of [*sic*] stayed with the men I was with but the T. V. A. Bought them out, and they haven't any other place for me. I owe them and it will take what I have got to pay them. So I won't be able to farm no more but it hadn't been for the T. V. A. This would not of [*sic*] happened."[18] In response, he received a form letter stating he'd be visited by a caseworker and thanking him for his concern.

From minor complaints to rudeness to threats made to caseworkers, some residents took caseworkers' visits as opportunities to unload their emotions and voice their displeasure with moving. Anger, frustration, and concern were common emotions expressed among residents. Joe Culbert was upset about removal, as TVA purchased part of his land, but not enough that he could use what was remaining to support his family. While Mr. Culbert was talking to a caseworker about his situation, his daughter appeared at the door and said, "Daddy, tell that man it's none of his damn business how old I am and that we are not asking the TVA for any work or anything and they don't need to know anything about us."[19] The caseworker responsible for visiting Jim Bolden had been warned that his family was "very hostile in their attitude toward the Authority and were most resentful of being asked detailed questions by so many different representatives of the Authority." John Andrews's family had been living in the same area "with very few contacts with people who live outside in the valley." The family did "not take well to the idea of having to leave their home."[20] James Shelton "emphatically stated he did not have sufficient resources to satisfactorily remove from his present home."[21] As a tenant farmer, this statement makes sense; he received no money for the move, nor did he have possessions or family members and friends who could assist him in relocation. Jim Mason told his caseworker he felt the TVA was doing him wrong by forcing him to leave.[22] Earl Legg was "disturbed" about having to move.[23] Will Rice did "not [feel] so good" about moving, as his family believed "they will be considerably damaged by their forced removal from their home this late in the crop season."[24] They ended up relocating 4 miles north of Guntersville, and the caseworker said of Rice, "He will be satisfied until the TVA makes him move again which he believes will be in the next few years, because he thinks they'll purchase the land he is currently on." The next time, he said, he intended to move to Texas, far away from the reaches of the TVA. These negative feelings toward the TVA had the potential to be problematic, as caseworkers noted that a bad "attitude, if shared by other members of the group, will certainly present a problem."[25]

Antagonism against the TVA was a legitimate threat against public acceptance. Emotions are contagious, and it was important that the positive emotions surrounding the TVA had a greater impact than the negative ones. During family removal, before the dams were finished and rural electricity was readily available, many families did not support the TVA's presence and also did not have a way to voice their displeasure. Signs all around them encouraged those who were antagonistic toward the TVA to change their attitudes. The fact that extensive surveys of the relocated were done to determine their feeling about the TVA is consistent with the idea that the TVA actively worked to minimize criticism and threats to its mission. Though the TVA could enact its powers as a government agency, it did not want to be seen as a bully, imposing its will upon the poor people of the valley. Instead, they needed to be able to make the argument that everyone in north Alabama was working together, as part of a grassroots effort, to change their lives, with the help of the TVA.

Among landowners, there were complaints that the TVA inadequately compensated them for their land, which also contributed to general unease, disappointment, and further "antagonism" toward the agency. Some felt robbed of land they felt was rightfully theirs, and no financial compensation could truly make them feel good about leaving their home. Despite the TVA offering what they determined was fair market value for owners' land, some were not pleased with the amount of cash they received. Landowner Mary Burnham told the caseworker she was not happy with the deal she got for her land, so much so that she became depressed over the situation.[26] Mrs. William M. Auston of Marshall County was dissatisfied with the amount the TVA offered for her land and "talked at great length of her misfortunes/ inconveniences of moving."[27] Mrs. Williams of the Guntersville area "was cordial" with the caseworker, "although she [was] very much disturbed over having to move so quickly and feels that her land has not been adequately appraised. She regrets the necessity of leaving her present home where they have lived for 12 years. This was the best home she ever had; it [would] be impossible for her to find another home with a well, pasture, garden, fruit trees as conveniently arranged as this one. She emphasized the fact that she was getting old and had just gotten her home fixed where she could live comfortably the rest of her life."[28] For this resident, relocation essentially meant she had to start over after years of working to make her home a place she could be comfortable for many years to come.

Repeatedly, newspaper articles labeled those who disagreed with or opposed the TVA "selfish," regardless of the reason. Particularly when it came to the land purchase efforts, reservoir area residents were reminded how fairly the TVA was treating them, not just in north Alabama but also in Tennessee, where nearby Norris Dam had recently been constructed. Local newspapers made it clear that residents should not argue with the dominant opinion that TVA was good, had good intentions for north Alabama, and meant good things for everyone in the area and for future generations. Newspapers constructed the message that resistance to TVA was futile, and those who may have opposed it were portrayed as outnumbered. Newspapers discussed the "overwhelming acceptance" of TVA from the beginning of the project.[29] Alabama Governor Bibb Graves said in one article, "it ill becomes any man of anything to place a stumbling block in the way" of the TVA.[30] And newspapers also wrote about public events celebrating the TVA, such as the Limestone County TVA Appreciation Day in July 1935, complete with "57 floats, gayly decorated and bearing banners affirming the region's loyalty to the TVA program, [which] gave color to the proceedings as the parade wound its way over a 25 mile route to Big Springs Park. Streets were lined with cheering thousands. Seven planes droned overhead as the first float arrived."[31] Who would want to argue with such a public display of affection for a large government entity, promising to bring drastic, positive changes to the entire area?

The *Redbook* magazine debacle of July 1935 offers an interesting case study of the amount of public chastising that was in store for those who opposed the TVA, and it demonstrates how paranoid the area was about losing out on the TVA's initiatives, for any reason. The July 1935 edition published a letter "alleged to have been written by one of the city's chief officials" that made "satirical reference to the city's contract with TVA."[32] The *Alabama Courier* said it was "very unfortunate that information of this character should appear as coming from Athens, where the people are 'sold' on the TVA. It is unfortunate that the power trusts, or some sinister influence, continues to pay lobbyists and writers to malign the first great project that has been sponsored in the south by the federal government. Neither Athens nor any other city that has tried the differences between the old-time methods and the TVA will ever consent to a change."[33] Fred Wall, the editor of the *Courier*, publicly "declared any resident who would cause the city to receive such unfavorable comment as that recently given this place was 'guilty of treason to Athens.'"[34] Soon after the first article detailing the *Redbook* coverage was published in Limestone newspapers, a meeting was held to publicly endorse the TVA, in an effort to further demonstrate that the citizens of Athens did not feel the way the author of the letter did.[35] Senator Hugo Black was asked to enter a resolution in the Congressional Record that expressed Athens's appreciation of the TVA.[36] The resolution was important to Athens residents because they felt failure to publicly acknowledge the TVA's assistance to their area would harm their relationship with the government agency.[37]

The timing of this incident is noteworthy, as the TVA was in the middle of a Supreme Court battle with the Alabama Power Company over the constitutionality of a government agency generating electricity, setting rates, and selling it to municipalities. Newspaper reports gave the perception that Athens would be in grave danger if the TVA suddenly turned its back on the first city in Alabama to receive power generated from TVA projects. Athens was also worried about the constitutionality of the TVA and paid extremely close attention to news of court decisions. Without cheap electricity, modernization was not likely to happen as quickly, if at all. Newspapers, citizens, and politicians all quickly helped fend off potential public relations disasters for the TVA.

Relocation was even more of a financial hardship for those who were already struggling to make ends meet. The entrenched poverty of tenant farmers and sharecroppers was nearly impossible to escape, and forced relocation worsened their financial problems. The TVA paid landowners for their land. But renters got no financial assistance with relocating. Any costs they incurred as a result of moving were theirs to pay. And, as relocation cost valuable time that could have been spent tending their crops, it was simply another hardship the nomadic families had to endure to survive.

Removal made finances more difficult for families in two ways. First, renters who moved to plots of land that were smaller and/or worse soil quality found it more difficult to turn a profit, however meager those profits might have been. This would have further negatively impacted renters' ability to pull themselves out of poverty and out of the sharecropping system that held them down. Rister Woods, J. D. Hammond, T. J. Adcock, and Albert Hargiss all complained about relocation and its negative effects on their meager incomes.[38] Dick Abernathy was particularly aggrieved, telling his caseworker he thought "TVA [was] ruining one of best farms in that section"[39] by purchasing the land on which he worked as a tenant farmer. There was land available on Sand Mountain, but it was "about the last place anyone would want to live"[40] due to rocky soil that would surely be difficult to cultivate.

Relocation negatively affected nonfarming families as well. Andrew Higdon's family lived in "deplorable" conditions, forced to live almost completely outside, and they had no income except what their eldest daughter earned at a Works Progress Administration–related job. The job, however, required her to walk 9 miles each day after relocation, which was a definite hardship.[41] Residents who fished for a living instead of farming or working in industry found themselves further disadvantaged, as the TVA's protective strip along the river rendered it impossible for them to continue fishing operations. Luke Flanagan was forced to move "four miles away from the Elk River and a fishing stand," making it extremely difficult for him to continue fishing for a living.[42] The caseworker noted in Flanagan's file that if he chose to take up farming, he would be better off in his new location, even though Flanagan had no other skills and no other way of working for income. Albert Bevins, who was living in "a miserable hut, patched and slung together," had to move twice due to his proximity to the Wheeler reservoir, which meant he ended up living far away from his mussel operation.[43] Living so far away from the river made this line of employment difficult, if not impossible, to maintain.

The TVA's purchase of over 230,000 acres in north Alabama created a land and housing shortage, making it difficult for families to find new places to live. Because the TVA claimed their goal was to help the people of the Tennessee Valley, they had some ideas for how to handle the social problems that might result from uprooting a marginalized population of largely unskilled, uneducated people. After failed attempts at Norris Dam to relocate people into one area or housing development, the TVA changed their plans for the relocated in north Alabama. They instead put the responsibility for moving and finding a new place to live on those who were relocated. Despite the care and concern of the caseworkers who directly connected with relocated families, the TVA failed to adequately plan for the long-term needs of displaced families.

Family removal directly impacted the troublesome housing situation in the reservoir areas. When it came time for day-laboring and tenant families to leave their ramshackle structures, they had a few options. If they were lucky, they could inhabit an already existing house or other structure on new land; sometimes, these structures were nothing more than shacks, similar to what they had left behind.[44] Sometimes they could rebuild their home with materials salvaged from their previous home; in many cases those structures had barely provided any shelter. If they lived in a house-boat, they could float it down the river. If they lived in a tent, the relocation would presumably be easier than tearing down and rebuilding a sturdier structure.[45] In the most extreme circumstances, the TVA supplied the family with a tent until they found something more permanent in which to live. It was difficult for housing conditions to get worse than they already were along the river. As one caseworker explained of one family, "the house from which these people moved was nothing but a shack. Any place they could have moved to would be as good as that from which they moved."[46] Apparently, the housing and land north Alabama residents moved on to did not matter as long as they moved off of TVA property.

The TVA bought some land with extremely poor housing only to rebuild better housing and small communities for workers. It is ironic that the TVA built better housing for their workers than that which previously existed for the residents. "Damtowns," as relocated resident Bill Hardin and Eugene Simonson, the son of a TVA worker, remembered them, sprung up in the shadow of the Wheeler, Pickwick, and Guntersville Dams, as well as other dams across Tennessee and Alabama. These towns, created for workers who moved to the area from other states, were vastly different than the housing of their poverty-stricken neighbors. Eugene Simonson was a child when his family moved to Guntersville from Arkansas so his father could work in the TVA's mosquito control division during the construction of Guntersville Dam. The life lived by the Simonson family, and presumably other families in the TVA camps, was much different from that of their new neighbors. Their housing was simple but cleaner and sturdier than the housing in the surrounding area. Compared with his home on a farm in Arkansas, life on the TVA compound was better. According to him, "it was a big step up. Because our farm in Arkansas was a subsistence kinda [sic] farm, we had no electricity there, and no indoor plumbing, of course. So moving here to Guntersville Dam was a big step up for us."[47] In fact, he said he "hated" their farm when they returned home to Arkansas after his father's time with the TVA. Mr. Simonson enjoyed many amenities in TVA housing, such as the community building, where they had an auditorium, a museum, movies, church services, and a radio. The fact that there was electricity for those who lived at the TVA camps, and very little in the surrounding area, is thought-provoking.

The TVA expected that providing comfortable housing for newly hired workers would result in greater productivity. TVA employees were treated well in these simple but comfortable villages.[48] The homes and dorms constructed in damtowns were modern and comfortable, especially compared to the dilapidated structures that surrounded them. Onsite cafeterias provided plenty of food for workers and their families. There were even leisure-time amenities such as a basketball court. In fact, Bill Hardin remembers the first time he saw a paved basketball court at the nearby TVA Guntersville damtown. Neither he nor his friends "had played basketball on a blacktop court before! TVA had basketball courts down there and a tennis court, and you know, we had never seen anything . . . on a level course like that." Before the TVA brought paved basketball courts to the area, the children played and practiced ball "with a hoop up on the side of the barn." For children who had never experienced the wonder of a paved basketball court, the TVA must have made exciting improvements to their lifestyle, providing something that children in more urban, populated, or moneyed areas took for granted.

Though Eugene Simonson and Bill Hardin were neighbors, they led vastly different lives. Bill had no choice but to work to help his farm family. Eugene, on the other hand, chose to work part time on a newspaper route and part time in the cotton fields because he wanted to do something for extra income for himself, not because he was forced to work to help his family feed themselves. Those in TVA damtowns were not "living in luxury," but compared to others in the community, they were of a higher social class, and had greater access to resources and opportunities.

In addition to housing problems, relocation often exacerbated considerable health problems. Families often could not afford medical treatment, making it necessary for caseworkers to coordinate with welfare agencies on behalf of the families they represented. In the most severe cases, caseworkers tried to arrange free medical care with benevolent doctors who would treat residents for free until they were well enough to relocate.[49] Malaria was common among the relocated, and it was not unusual for entire families to suffer from the illness, such as reservoir area resident John P. Dunn, his wife, and three of their five children—all of whom had malaria at the time of the caseworker's visit.[50] Joe Blackwood's entire family, all living in a one-room house,[51] had malaria; Bess Keeton Hooper's family and J. A. Keasling "all had chronic malaria."[52] Walter Allen in Jackson County had "occasional chills" like "other residents of the Goose Pond Community."[53] Malaria made moving even more difficult, as it took strength and energy to move.

In addition to malaria, residents sometimes needed medical attention due to other illnesses. Thomas Green could not move at the time his caseworker visited him because he was too sick to leave his current home.[54] Living on a houseboat "ruined" the health of Mrs. John Langston, such that she probably never physically recovered after family removal.[55] Vernon Sheffield died from the tuberculosis that plagued him before the caseworkers launched a lengthy attempt to relocate him.[56] Children were not immune to illness either. Dave Rice Jr.'s baby died the afternoon before his caseworker visited the family, after suffering 2 weeks with diarrhea. Their caseworker noted "everything about the place gave testimony of poverty."[57]

Aware that their priority was to relocate residents, caseworkers were still sympathetic to these impoverished people, especially when it came to dealing with the extremely sick.[58] They did what they could to help.

The case of James Perry Bowling, who lived with his wife and five children on Gilchrist Island near Lawrence County in the Wheeler reservoir area, is another example of how poor health impacted entire families. Prior to being removed from their home by a TVA boat, Bowling worked as a farm laborer in the Wheeler area for $1 per day, worked on another farm near Guntersville, and worked a short stint in the textile mills in Huntsville.[59] Both Mr. and Mrs. Bowling suffered from health issues, as indicated by over sixty pages of documentation about their health and related relocation problems. TVA's Wheeler Dam Assistant Medical Officer Walter T. Davis described Mrs. Bowling as "an extremely emaciated, malnourished, anemic individual who complains bitterly of heart attacks and chills" and he suspected that she and the entire family suffered from malaria. She also had scurvy, and as the doctor noted, "the most severe case of pyorrhea that I have ever seen." The children were unhealthy, too. Their caseworker said that "two of the four children show some evidence of beginning scurvy as well as much malnutrition and anemia."[60] Additionally, "the whole family need[ed] dental attention very badly." The Bowlings lived in a tent lent to them by the TVA. However, during his time in the tent, Mr. Bowling shot one of his eyes out while hunting rabbits. Doctors told him to get the eye removed, but he could not afford the surgery and failed to follow their orders. This understandably led to complications. TVA caseworkers eventually found a doctor in the area who would do the surgery for free, as long as the TVA agreed to find Mr. Bowling and his family a house in which to live, rather than the "undesirable" tent they were inhabiting. TVA caseworkers also took Mr. Bowling for follow-up medical care for 2 weeks, demonstrating an example of the level of compassion they felt toward the relocated in the most destitute of circumstances.[61]

The health of relocated residents was not discussed in newspapers, nor was the complicated issue the caseworkers faced of getting adequate medical treatment for those in need before they could feasibly move from their land. Though the TVA's ultimate goal was to improve the health of the entire region, for a considerable number of people, relocation made their health worse before it got better. The TVA's publicity efforts focused understandably on how health would be improved and gradually incorporated messaging that showed how much healthier the entire region was after the TVA's influence. Here again is an example of a situation in which newspapers could have written about relocation in a way that framed stories within a pro-TVA narrative. Caseworkers had a direct hand in relocation, which was unpopular, but they also were heroes for many people who were not receiving adequate medical attention. Stories about the work they did could have been framed as another way that the TVA was helping the rural residents who needed better health care, but newspapers did not cover this angle.

Finances, housing, and health were not the only factors complicating family removal. One's reputation in the community played a major role in whether or not one's relocation was successful, particularly for tenant farmers and sharecroppers, who depended on landowners for land to farm. Landowners were reluctant to take on tenants who had a reputation for failure to work or drinking, or who were otherwise known as troublemakers. Caseworkers noted these concerns in numerous files, especially for families in the Wheeler reservoir area.

Whiskey production and consumption was a popular activity, as Bill Hardin revealed in his interview: "Everybody back in that time was either farmers or bootleggers, and sometimes both, and that's the way it was. It was a way of life back then."[62] Liquor production should be understood as a source of income for some. One of Elias Fitchead's neighbors told a caseworker that "Elias has always been trifling, but that it is said he always made a good living making liquor."[63] Still, with the reduction in the amount of land available for corn crops and the forced removal of individuals from the backwoods into more populated areas, family relocation had an impact on this part of life in rural north Alabama as well. Several Knight's Island residents were known for making and selling whiskey, not all that surprising given the only thing grown on the island was corn. Hilton Malone, for example, would probably have found it hard to relocate because "he drinks whiskey to excess," as would his neighbor, John Baty, who lived on a towboat owned by Mr. Knight. The caseworker suspected he was probably a "moonshiner" given the amount of corn he grew for personal use.[64] Mr. Scoggin, who also had a reputation for moonshining, would likely run into issues finding a new place to live.[65] Caseworkers certainly made assumptions about some of the relocated, but sometimes landowners or neighbors suggested personality traits that they thought would be problematic for relocations. For example, one landowner said of a certain tenant: "Tom is a tin-can moonshiner and a very undesirable tenant."[66]Reputation could determine whether or not an existing community would welcome an individual or family into their ranks, or work hard to keep them away, and the reputation of a family tended to follow them wherever they went. Such was the case for one north Alabama family. Their caseworker said the community never accepted them, as "everyone seemed inclined to make some slight remark about how poor they were or how 'good for nothing' Mr. A was."[67]

Most families were hard-working, though, and did their best to survive their poverty-stricken conditions while not bothering neighbors. For example, Waymond and Hattie Hapton lived with their eight children in a three-room house in poor condition on the land owned by Mr. McGehee, who happened to be neighbors with Frank and John Mack Conner. Waymond had been farming for 45 years, since he was 10 years old. In 1938, at age 55, Waymond and his family, like their neighbors, the Conners, had to move from what became TVA property. The Haptons managed a fairly large "3 mule crop" of cotton and corn. Their worldly possessions amounted to furniture, two cows, thirty chickens, two hogs, and one hundred cans of fruits and vegetables.[68] Paul Conner fondly remembered the Haptons. In fact, Paul said, "I called him Uncle Waymond, and his wife was 'Aunt Ad.' She worked for my grandmother. And he was a syrup maker."[69] There was one big difference between the Conners and the Haptons, however. The Haptons were black. To the Conner family, the Haptons were simply part of the community who helped tend crops and maintain the farm; the Conner family depended on the Haptons just as they did other sharecroppers and tenants.

But Waymond Hapton's case file revealed that not everyone was similarly disposed to the family as Paul Conner. Relocation was harder on the Haptons because "farming opportunities for colored people [were] very limited in [the] area."[70] As of September 15, 1938, they had rented a farm near Fayetteville, Tennessee, but a later statement shows the family moved 5 miles away into Southtown, an area where many black families eventually relocated. The family was better off in terms of their proximity to stores, schools, and churches, but there were no farming opportunities in Southtown. For a family who had known nothing but farming their entire life, relocating to an area where farming was impossible would have made it difficult to survive. Waymond's plan, according to his files, was to move to the farm purchased by Mr. McGehee in Tennessee to resume farming in 1940.

Blacks were further disenfranchised from the land acquisition process because, as nonlandowners, they did not receive payment for land, and they did not receive relocation assistance. In the Wheeler reservoir area, only 7 percent of landowners were black.[71] And in the Guntersville reservoir area, only 1 percent, or 18 out of 1,182 families, were owners.[72] Blacks were not always allowed to live in TVA damtowns, such as Norris Dam.[73] Other damtowns, while they let in blacks, were deeply segregated. Residents in white-dominated areas often refused to let blacks live nearby. Black residents who ended up leaving farms to move to urban areas may have faced even more serious segregation.[74]

Blacks and whites were relatively friendly neighbors in many cases; in others, segregation ran deep. It was difficult for some black families to find employment and places to live when the TVA purchased land for the creation of reservoir areas. And, despite their commitment to hire black workers, the TVA did not completely deal with the unique problems facing black families who were forced to relocate. Though black farmers "were looking for and hoping for a revitalization program like TVA, which promised better treatment and living conditions for everyone in the valley," the TVA did not always deliver for them.

Racist attitudes made relocation for black families extremely difficult. Deep in the mid-1930s segregated South, racial tensions were exacerbated by the stress of relocation. The simple fact that a family was black could present a problem. One caseworker stated that "farming opportunities were limited for colored people,"[75] due to the land shortage and difficulty of finding owners who would rent land to them. A similar sentiment was echoed in Robert Hayes's file, as the caseworker noted the only problem with his removal was "finding him a place where he will be able to work next year, as places for colored families in this area may not be plentiful."[76] The caseworker tried to get Robert interested in relocating to Scuppernong Farms in North Carolina, but he declined to even go see the place. Eventually, he decided to remain a tenant for the man for whom he had worked for many years.

Black families were most often poorer than their white counterparts, and they had the additional hardship of dealing with segregation and racism. For example, Clarence and Maggie Points were black landowners who had lived in the Flower Hill section of Lawrence County for 24 years. Clarence, a "helpless cripple" who "lost the use of his legs about three years" before the TVA began building in the area, believed he owned the house and 200 acres of land on which it sat, "until the T.V.A. bought up the property in the Reservoir."[77] Clarence could not prove ownership and subsequently lost all the money he had invested in it, which was about $4,500. This was, understandably, "a big drawback to both of them." They were left with just $763 worth of total assets, including furniture, three mules, two cows, two hogs, and a wagon worth $10. Their caseworker noted, "Though it is not clear why the Points family was allowed to live on the same, large acreage of land for so long without a deed and without incident, it could have been due to the local custom of nontitled land as a means for supplementing income."[78] The TVA's refusal to pay for their land was a significant hardship, placing this black family in an even more vulnerable position than their white neighbors. The Points family was forced into a lower class of tenancy or sharecropping—demoralized at best, and cheated at worst, by the entire process.

The lack of housing available for relocated black families was a particularly severe problem. Ernest and Lucy Horton had four children, who at the time of caseworker Martha Branscombe's visit were "all in rags and filthly [sic]."[79] They lived together in a "two room, frame building in dilapidated condition. There is a small front porch. The walls, unceiled, were covered with pieces of corrugated boxes to fill up the cracks and gaps. The roof is no protection against the rain. The house is poorly furnished. . . . Everything was dirty, cluttered with rubbish and swarming flies." Ernest, who was a veteran and had been a farmer, was unable to work due to "heart or lung trouble," so Lucy was the sole breadwinner. Her income came exclusively from washing clothes for four families. Branscombe stereotypically described Lucy as "a rather heavy set woman with all the typical racial characteristics of the negroes. She is of a jovial disposition, talks continuously in a blustering voice and laughs loudly and often."[80]

Branscombe came face-to-face with the housing problem when she tried to find lodging for the Hortons in Southtown. She talked with local businesswoman Cora Carter, who owned multiple homes in Southtown, about the possibility of renting out some of her homes to the black relocating families, who desperately needed a place to live, but Mrs. Carter was unmoved. Branscombe talked with others in the community, including the mayor of Guntersville and a Willie Saunders from nearby Albertville. Saunders encouraged Branscombe to help black families move to Albertville, where there were more opportunities for housing and employment. However, Albertville was 10 miles from the Guntersville city center, which would have resulted in a considerable move for families who had no resources to help them move that far.

Henry Horton and his wife, Betsy, lived across the railroad tracks from Ernest and Lucy Horton. Their removal situation was similar, although Ernest had difficulty finding a job. Caseworker Branscombe feared that Ernest and Lucy would likely have problems moving because they were part of "a sub-marginal group and due to the scarcity of houses, this family will likely present removal problems."[81] Branscombe eventually talked with Mayor Couch and other community leaders about "plans of the city to make provisions for colored families." The Hortons all considered moving to Albertville, but because there were few employment opportunities for women in the city, they decided to look elsewhere. The family tried diligently to find a place to move until Cora Carter constructed a new home for the them on her property that had "two rooms, is screened, has running water, paned glass windows and a toilet. There is adequate space for a garden. The house is . . . much more healthy than the former one."[82] The Hortons were "elated" with their new home. Both Ernest and Lucy Horton went to work for Carter: Lucy cooked for $5 per week, and Ernest did laundry for $2 per week.

Another example of the relocation problems black residents faced was Bud Merrill, who should not have had any more difficulties relocating than his white neighbors. But "considering the fact that colored people are not wanted in many communities of Alabama, it may be very difficult for this family to rent as much farm land as they now cultivate."[83] Similarly, L. W. Scott felt it would be impossible to "purchase land within a reasonable distance equal to that sold to the authority without paying a much higher price. This, he feels, will be especially true inasmuch as many communities are closed to negroes."[84] Henry Chandler, a former slave, was "forced off the block"[85] in which he lived peacefully and died almost immediately after moving. It was difficult for well-meaning residents to help black families due to the widespread racism in southern communities. Black families who were relocated had no choice but to submit to the TVA's demands, even though doing so might stir the ire of racist neighbors.

Stereotypes and racist assumptions about black residents were reflected in some of the caseworkers' dealings with their clients. One caseworker stereotypically described Ernest Ruffin as "more industrious than the common run of negro families."[86] Another caseworker noted that I. C. Griffin "had the usual philosophy of life that is shared by most old negroes, that of just waiting to die and yet a little afraid to die for fear that he had not lived hardly as well as he could have."[87] Though the TVA made attempts to desegregate its workforce, the agency as a whole was criticized for its policies on race.[88] No special plans were in place for how to deal with displaced black residents, citizens who were further marginalized beyond their socioeconomic status. The TVA made no special concessions or efforts to ensure black families were well situated after their removal from the reservoir areas. Instead, it was up to the families, most of whom received no financial compensation for moving, to fend for themselves, outnumbered in an area dominated by white, sometimes racist, families.

Newspapers largely ignored this important social issue. In the segregationist South, the disproportionate struggles of black families would not necessarily have been newsworthy. It was just the way life was. Articles praising whites and blacks for working together, as we saw with Cora Carter and some of the relocated black residents of Guntersville, would not likely have been popular in mainstream north Alabama newspapers. As a result, the complicating factor of race went unnoticed among those who were keeping up with TVA news.

Where did families go when the TVA told them they had to move? After living in the same place for 40 years, what other location would serve as a decent option? What do you do when there are no houses for you to move into, either because they do not exist or because you are black and the white people block you from their neighborhoods? Families who grappled with these questions must have faced fear, grief, anxiety, anger, and frustration, particularly because they had no choice but to move. And local north Alabama newspapers—under the influence of editors who were vocal TVA supporters and who readily publicized information created by the TVA's Information Office—did not fully address the complicated, extreme, and varied difficulties facing relocated families, black or white. Instead, newspapers and other media messages emphasized the improvements the TVA promised, the number of jobs brought to the region thanks to construction of the dams, and the overall benefits of soon-to-arrive rural electricity. This helped cultivate a positive public opinion of the New Deal agency.

The TVA had extensive, detailed, and well-designed plans for a changing geography and economy, but they did not have large-scale, long-term plans to serve the relocated. The TVA's primary objective was that the residents move elsewhere. The agency's philosophy was that it bore no real responsibility for finding residents a new place to live, which presented a significant problem, especially for those who depended on landowners with large farms for support. In north Alabama, there were already housing problems before TVA building programs began. In the Guntersville reservoir area, there was not "a single habitable vacant house" when the TVA started purchasing land,[89] which meant trouble for the 1,182 families who had to find somewhere to live. Nearby Huntsville, Scottsboro, and Albertville had scarce housing facilities, so Guntersville families had difficulty moving to those towns.[90] Relocating the population in the Wheeler reservoir area presented a different set of problems because most of them were nonlandowning farmers. Although many relocated families moved frequently because of the nature of the tenant system, they rarely moved far from their previous location. The TVA recognized that their main task in this reservoir area was helping those who did not own farms to move away from farming into other occupations.[91] Unfortunately, there were few other viable employment options for nonfarmers in that area due to a lack of industrial development and poor economy. Nonowners essentially were left to take up residence wherever they could, which meant many were forced away from farming. These issues were rarely discussed in north Alabama newspapers.

Despite the lack of land and housing options, most relocated residents remained close to their original homes. In the Wheeler reservoir area, 89 percent of displaced families stayed within their original county of residence, while the rest moved to another county or out of the state.[92] The same was true in Guntersville and Pickwick, in which 93 percent and 95 percent of residents, respectively, stayed in their home counties.[93] As the TVA expected would happen, most relocated families did not go very far from their original homesteads. But the twin problems of finding adequate housing for the majority of relocated residents who opted to stay in the immediate vicinity and the loss of over 200,000 acres of TVA-purchased land were daunting, to say the least.

Not surprisingly, social and economic status played a significant role in relocating options. Tenant farmers, sharecroppers, squatters, and the unhealthy or elderly faced considerable difficulties finding new homes. Caseworkers tried to relocate less fortunate individuals into their relatives' homes, but moving in with relatives in already-cramped houses tended to add further complications. Some relocated families quite simply did not know where to go, which was a major issue for the caseworkers who were responsible for making them move. Mrs. Joe Hutcheson admitted "she did not have any idea where she could move that she was sure they would not cause the TVA any trouble and would be moved by [the TVA imposed deadline of] January 1."[94] Sometimes, difficulty in finding a new place to live caused delays in relocation.[95] Nonfarming families ran the risk of losing jobs, such as B. L. Hester of Marshall County, who had to move just a half-mile away from his previous location but lost his job operating the Town Creek Ferry because a ferry was no longer needed after the TVA came into his community.[96] Others were faced with having to move too far away from their existing jobs, such as Cleve Edmonds, also of Marshall County. Edmonds hesitated to move his two daughters who worked at the Saratoga Cotton Mills in Guntersville because "their removal would take them away from their work unless they boarded in town."[97] Comparatively fortunate landowners, who had relocation choices and were in good health and able to work, found the decision about where to move was heartbreaking nonetheless. Paul Conner remembered his father making that difficult decision, calling it "agonizing." His father and his uncle hoped to buy new plots of land close together, as they had been before the TVA moved in. They looked at land in Huntsville and on Sand Mountain. The night before their decision, they stayed up all night debating where it would be best to live. Eventually, they decided to take their chances with Sand Mountain.[98] Regardless of wealth, social status, or age, relocating was a painful process that was repeated thousands of times in north Alabama.

Understanding that there were few options available to relocated farm families, the TVA encouraged Guntersville farm families like the Conners to relocate to Skyline Farms in nearby Jackson County, Alabama, Scuppernong Farms in Columbia, North Carolina, or the Wolf-Pitt Project in Rockingham, North Carolina. These developments were government-owned farms designed to give low-income farming families a place to live and work. The TVA promoted Scuppernong in a letter, saying "the Farm Security Administration owns about 16,000 acres of land in North Carolina. Each family participating will be furnished with a house and garden."[99] The thought of new housing and a fresh start in an area far away from the TVA's presence must have sounded promising, even though the thought of moving far from home was likely intimidating. For those in the Guntersville reservoir area who did not want to move so far away, Skyline Farms must have been a more attractive option, given its close proximity to their existing homes. However, a move to any of those areas was risky, and not everyone was enthusiastic about relocating there, regardless of the promise of adequate farmland, because soil quality was so uncertain.[100]

Soil quality aside, a place at these farms was not offered to everyone. Families had to apply and be screened. They were considered for relocation to Skyline Farms only if they demonstrated a desire to reestablish themselves, possessed an ability for management, and a cooperative temperament; men had to produce references who would vouch that they were good workers and good men.[101] Whether someone was suitable for relocation to a farm was determined by caseworkers and feedback from references. Oscar Elledge was found to be qualified for Skyline by caseworkers because "1. He is evidently a good farmer. 2. He has ambitions to own a home. 3. This is a family of young and hard working people. 4. He has accumulated enough livestock and other resources so that he would be able to make a start by himself. [Further, other factors considered included] 'age—23 years, farming ability, well recommended by present landlord, cooperative attitude of family, in need of such aid, apparent good health.'"[102] Farmer Bob McCrary was not so fortunate. He was interested in Skyline Farms, but his caseworker noted, "The following facts speak for themselves: 1. The man's age is not favorable to his being able to complete a long-time program of payments. 2. References indicate that the man and his family are not good workers. 3. General appearance of home indicates that proper care is not given to things provided by the husband."[103] According to the TVA readjustment report, only a small percentage of north Alabama residents moved to the government-owned farms. Skyline Farms accepted six families from Guntersville, Scuppernong Farms accepted sixteen families, and the Wolf-Pitt Project accepted one family.[104] As with any place families could have chosen, the farms were segregated, presenting one less option to black families. Black families were generally expected to make their own arrangements.

Farmers who managed to find enough land to continue farming faced a second problem—finding sufficient land that viable enough to grow crops. The TVA assisted farmers through agreements with county extension agents at land grant colleges, such as the Alabama Polytechnic Institute (now Auburn University). Extension agents helped relocated families find places for rent or sale and worked with local farmers to develop better farming techniques.[105] The TVA saw this as a positive offering; one TVA publication noted "although the average farm acreage has decreased somewhat among farm families . . . productivity has increased because of improved farm practices encourage by the Extension Services."[106] Some farm families still struggled though. At Skyline Farms, keeping the plateau land fertile enough to produce a worthwhile crop was challenging. Herman Baker, a Guntersville reservoir resident who moved to Skyline, eventually told the TVA that he had to leave because he could not grow crops there.[107]

Relocated farmers had a difficult time moving to new land and using new techniques, but additional readjustment problems faced farmers. Nancie Nelson found it "hard to adjust" to her new home, as she was forced to move close to a railroad and "had difficulties with people stealing chickens, hogs and a cow."[108] Alex Elrod was not satisfied with his new location, either; he was left with no garden and no income.[109] John Moore was "antagonistic" because his housing conditions did not improve after he relocated, and he eventually lived in a shack instead of a house; he had lived in a house before relocating.[110]

Local newspapers all but ignored the plight of relocated families. They did not report on the heightened activity along the river, the extra hardships relocating families faced, or the emotional trauma inherent in forced removal. Local newspapers also did not report on housing or land shortages or on the changing economies. They failed to report many significant stories including the racial tensions exaggerated by relocation, the necessity for farmers to scale back operations due to land shortages, overcrowded housing conditions, and the TVA's limitations regarding direct assistance to displaced farmers and their families. However, these issues were both newsworthy and worthy of critical attention. Such stories may have generated negative press for the TVA, something the agency wanted to avoid. TVA needed local communities to accept their presence and work with them in order for the agency to be successful.

Local newspapers helped the TVA achieve this goal by touting the long-term economic benefits and other positive aspects the agency while choosing not to cover contentious issues. They extensively reported on the modernization from which its readers and advertisers would benefit. They did not, however, balance their coverage with the many sacrifices of homes, families, land, and livelihood. By focusing only on positive aspects and ignoring the negative impacts of TVA electrification, the press played a fundamental role in shaping the story—and the history—of the TVA in north Alabama.

Despite their hard feelings toward the TVA, residents in north Alabama moved peacefully, largely without vocal or legal resistance. Ultimately, any resistance fueled by negative feelings, disappointment at leaving, or significant social problems was futile. No one was able to stand in the way of dam construction or could avoid forced relocation to other areas of the county, state, or country. Furthermore, newspapers did not communicate about the forced relocations, and there was no public forum for those who were relocating to express grievances.

Local north Alabama newspapers constructed a skewed reality for their readers. Though newspapers printed accurate information about the TVA, in omitting coverage about the thousands forced to relocate to make way for TVA projects, they failed to accurately communicate reality for everyone in their readership area. Newspapers shaped public opinion by printing stories that emphasized the promise of economic growth and modernization much more than turning a critical eye to poor people's hardships and their need for greater assistance. For poor farming families, the benefits touted in local newspapers were not experienced immediately. These families dealt with severe changes to their way of life and had very little assistance from the TVA. Caseworkers helped as much as they could by coordinating with other local, state, and federal relief agencies, but it simply was not enough to truly assist people in raising their socioeconomic status. Pulling an entire region out of poverty took much longer than newspapers would have had readers believe. Among the successes were failures, too, in which newspapers missed opportunities to write critically about the plight of the farmers, and the TVA missed opportunities to engage in large-scale social welfare planning for those who were directly affected by family removal. These two missed opportunities further disadvantaged those who sacrificed the most in the process of modernizing the north Alabama.

Conclusion

Luther and Beryl Tidwell remember clearly what it was like to get electricity for the first time. So fascinated with the new light bulb that illuminated their house, having never seen anything like it before, Luther said he and his siblings would "just stand there and pull the string" that turned the light on and off.[1] They were not quite sure how to react when their modest home suddenly featured this luxury, which was already taken for granted in many parts of the United States. It was exciting to see power trucks raising utility poles across fields in north Alabama. Rural electricity was arguably the TVA's main selling point to the residents of the Tennessee Valley. North Alabama residents needed affordable electricity to help modernize their homes, farms, and lives, as they were slower to obtain electricity compared to the rest of the state and the rest of the country. It was but one component in the TVA's grand plans to modernize the South, a government experiment that, if successful, might be replicated around the world.

Measuring the TVA's success is a difficult task. TVA annual reports extol the number of homes modernized by electricity, the number of jobs created by dam construction and reservoir clearance, improved navigability of the Tennessee River, and resulting social and economic improvements across the region. These are accomplishments for which the TVA is worthy of praise: they contributed substantially to the modernization of the South and for ending the economically harmful sharecropper system. However, the TVA's self-produced literature did not address other important issues, such as the emotional turmoil over leaving ancestral homes, the separation of families, or the extenuation of hard times before real economic improvement reached the rural areas. Little was said publicly by the TVA about what became of those families, the ones that caseworkers visited five, six, fifteen, sometimes twenty times to convince to leave. Though the TVA did some follow-up studies on a sample of families who relocated, they did not attempt to follow the lives of all relocated residents. As a result, not much is known about the families who were thankful for TVA-provided tents, which often provided more shelter than the shacks they had lived in before relocation, or families whose houseboats drifted downstream, hopefully to calmer waters. These families' stories tell us that, even when TVA provided financial resources to move, the process was difficult, uncertain, and emotionally draining. TVA family removal procedures reflect the TVA's lack of a comprehensive social plan for those who were affected by dam construction, the ways in which mediated and interpersonal messages supported the TVA as opposed to dealing with critical thoughts and actions, and the role of a public relations office in shaping public opinion.

Despite the TVA's detailed records regarding relocated families, later review reveals inconsistencies in the reporting of attitudes among families who moved, as family case file narratives and later reports of those who moved do not always match up. A few explanations for this phenomenon are possible. It may have been difficult, if not intimidating, for some residents to honestly tell a TVA representative that they were opposed to its programs and presence in the Tennessee Valley, for fear of retribution or simply of starting an argument with a government employee. When a representative of a powerful government agency is standing on your front porch and that representative asks what you think about the agency that will dismantle your home and flood your property, the answer must have been more complicated than a survey question could capture. Hazel Moore Thompson's family was not as enthusiastic about moving as her family's case file said they were. Luther Tidwell's family readily cooperated with the TVA's demands, even though they hated to leave their home and the move caused his mother to sink into depression. His family did not tell their TVA caseworker about their displeasure with having to move, nor did the family of Maxine Williamson Black.

Whatever their feelings toward the TVA, it's clear that in north Alabama, almost no one publicly spoke out against the agency. This made it look as if *everyone* favored the New Deal agency. Newspaper coverage was just one way relocated families were reminded that if they were against moving, they were in the minority. Who would dare to argue with anyone who could control electricity, as evidenced by the TVA logo depicting a fist holding a lightning bolt? Which hard-working farmers, barely able to feed their families, would oppose an organization that brought them, and their neighbors, jobs and money? Who would resist affordable electricity that promised to make their lives easier? Who would deny the South needed a change that seemingly only a powerful agency backed by the federal government could deliver: an escape from poverty, inadequate housing, health problems, and an agriculture-based economy that held many farmers hostage in a seemingly unbreakable sharecropper system? Even the Supreme Court agreed that the TVA's programs, despite their extensiveness and initial disruptiveness, were constitutional. As a result, there was a quiet and mostly peaceful acquiescence to the government's demands for land and relocation.

Feelings of antagonism toward the TVA caused some to be reluctant to move or to engage in other resistance strategies. Reluctance to move was arguably the best resistance strategy residents had; almost no one had the financial resources to hire a lawyer to fight their removal or the government purchase of their land. Perhaps those residents thought that if they stood their ground, the TVA would simply go away. In some instances, caseworkers made over twenty visits to a home, illustrating the quiet determination of many families. Though the TVA had the ability to use legal action to forcefully evict residents from their homes, they never used it in north Alabama. Eventually, everyone complied with the TVA's demands.

Relocated families had very little power in this process. They were told to leave by a government agency that claimed to be looking out for their best interests, even though it may have appeared they were being further disadvantaged. Because people trusted the government at that time, many were able to put aside their personal feelings about relocation because they trusted the government to provide a better way of life for them and future generations. In addition, despite any personal feelings they may have had, residents felt an obligation to obey the law. A TVA caseworker telling residents that they were trespassing on government property was an effective persuasive strategy.[2] Bill Hardin remembered people talking about resisting relocation, but when it came down to it, "most people were law abiding citizens back then and if the officials of the county said it's out of our hands, most people just said, ok, we'll do it then."[3] Maxine Williamson Black remembered that her father and many "of the farmers through here got really upset about TVA coming in taking all the land. And my father said, 'well, you can't fight anything that big, in that government. So the best thing to do is just to see if they can help us get this hill country where we can make a living on it.'"[4] This trusting attitude toward the government worked in the TVA's favor, as it meant most families moved without much resistance.

Those most directly affected by the TVA's construction in north Alabama were accustomed to living in poverty, with frequent health problems, and in inadequate housing. As Beryl Tidwell put it, "Nobody had anything back then. You didn't want anything 'cause you knew you couldn't have it."[5] They were trapped by a sharecropping system that was difficult, if not impossible, to break, mostly farming land that was overworked and unproductive. But the land along the riverbank that was taken for the creation of reservoir areas was fertile. Those who relocated were largely uneducated due to economic and circumstances beyond their control, with no power to question authority. This does not mean relocated families were ignorant. They fully understood what was happening. They simply lacked resources to resist. Most did as they were told, because they felt they had no choice. The TVA's plans were "accomplished with a minimum of political domination, control, or interference,"[6] in part because of the acquiescence of north Alabama river families. The relocated had few avenues for stopping the government-forced relocation or for willing the newspapers to cover negative impacts of the TVA program. The media also were of no assistance to those who felt disadvantaged, as local newspapers opted to support the government agency. Had relocated families pursued legal action, not only would they have had to fight the TVA, but they also would have had to fight a dominant ideology that told them they were wrong for opposing it—an ideology that the Information Office directly influenced. Newspapers and neighbors influenced by the TVA's publicity efforts reminded readers that TVA was helping their area, and it would be foolish to disagree with them. Speaking out against their neighbors might have been a tougher battle than pursuing legal action against the government.

But perhaps the strongest force keeping relocated families from actively protesting or pursuing legal action against the TVA was the promise of improvements. Despite their own personal hardships, it was clear that most of the relocated felt a sense of doing what was right for the greater good. Many were optimistic about the potential benefits of TVA programs, as it seemed like the last resort for help in the Tennessee Valley.

In her book, Marguerite Owen explores some of the ways in which rural southerners' lives improved after the TVA came to the Tennessee Valley. She wrote of numerous social, economic, and health improvements attributed to the TVA and the immeasurable "enhancement of serenity, [or] increased joy in life."[7] But quantifying the TVA's impact is difficult. One common metric of success is to look at the program's return on the financial investment. The cost of the TVA program was not an issue to the federal government. In fact, the more money the TVA spent in the area, the more important the dams seemed. By 1939, the TVA had purchased over 263,000 acres of land.[8] The construction of Wheeler Dam cost $30.3 million, Guntersville cost $33 million, and Pickwick cost $45 million.[9] By the end of the TVA's construction efforts in north Alabama, two large concrete dams stood in the middle of the Tennessee River, at the eastern and western ends of the state, controlling the flow of the once-unruly waterway, making it navigable and harnessing its flow to generate power for an entire region. To many, it was viewed as money well spent; it was money that put desperate

Americans back to work and modernized the South. Moreover, the economic impact in the local communities the TVA served was significant, as municipalities benefited from the sale of electricity through cooperatives. Local businesses enjoyed increased profits in selling more electric appliances. Those same businesses had renewed reasons to buy advertisements in local newspapers, reminding readers that purchasing electric appliances was yet another way they could support the helpful TVA.

The TVA promised and delivered many changes in the geography and economy for the people of north Alabama. Farming was still the predominant industry across the state, though Alabama's farming population dropped from 50.6 percent in the 1930 census to 39.6 percent in the 1940 census. The TVA had an undeniable impact on the presence of electricity in rural, areas in Alabama, especially on farms. In 1930, only 2.5 percent of Alabama farms had electricity; by 1940, 15.4 percent did.[10] People who were accustomed to washing clothes in the creek, having ice only as a luxury, and living by the light of small kerosene lamps eventually were able to purchase washing machines and refrigerators and turn the lights on, powered by electric current with just the flip of a switch. The TVA helped the economy by giving jobs to hundreds of north Alabama residents, mostly doing manual labor, clearing the land of signs of the past, helping to repurpose it for the future. Malaria rates decreased due to efforts focused on controlling the rampant mosquito population through pesticides. The Tennessee River became a predictably navigable river, as the dams corrected river depth around Muscle Shoals.

Forced out of the backwoods and closer to centers of activity in a county or city, relocated families moved nearer to schools, churches, stores, and health care. In a TVA-conducted postremoval study done on a sample of one hundred families in the Wheeler reservoir area, "67 percent of the former reservoir residents showed favorable attitudes toward new locations and to the Authority."[11] Only 7 percent of relocated families moved to "less desirable locations"; 39 percent of families improved their locations while 54 percent remained the same.[12]

Desirability was determined by access to the community, schools, markets, and ability to find employment. However, there are two issues with these statistics. The first is that TVA did not state how they drew their hundred-family sample from the more than 800 families who had to relocate in the Wheeler area; it is possible that sampling bias affected the resulting conclusions from that report. The second is that a better way of life involves more factors than simply the "desirability" of one's location. Did removal change anyone's status from owner to renter, or vice versa? According to the final report for families in the Wheeler area, there were 62 landowners prior to the introduction of TVA programs and 69 owners afterward, and there were 776 tenants before removal and 752 afterward.[13] However, just because a tenant became an owner, or because a person opted to abandon farming for another type of employment, did not mean his life drastically improved. Sam Boldin, for example, became a landowner instead of a renter after his forced relocation, but TVA caseworkers noted he was not as well off as he was prior to the move. His new farm, located at the base of a small mountain outside Huntsville, did "not compare in fertility to that of his former location."[14] He would likely have a difficult time growing crops on the land he owned. Tom Hardin faced similar hardships, as he moved his family from a farm with fertile, riverbottom soil to a plot with rocky, mountainous terrain. There was a learning curve associated with learning how to effectively work such land.[15] Boldin and Hardin were not the only ones left needy after the TVA forced them to move. After relocation, 179 farm families needed some aid postremoval that the TVA did not provide or coordinate.[16] Relocating always brought the possibility of tenant farmers "being unable to retire their indebtedness, which would, of course, force them back into sharecropping." Census data suggests homeownership, at least statewide, improved; the number of families owning their home went up 2.5 percent across Alabama from 1930 to 1940.[17] However, 64 percent of Alabama's rural farm residents were still tenants, and homeownership actually decreased in the TVA reservoir areas during the 10-year period. As table 4 shows, the 1940 census demonstrates a decline in housing for almost all counties impacted by TVA reservoirs in north Alabama.

Even when owners were not financially compensated for their lands, forced removal sometimes resulted in better conditions for families. Pleas Orr of the Wheeler reservoir believed his life improved because his new home was "comfortable," much better than his previous home that was "very cold during the winter."[18] Herbert Pack was also pleased with removal because he moved from living in a swamp to dry farm land.[19] Over time, health and sanitation conditions in the region improved, in large part due to the TVA's malaria control efforts. Only 9 percent of families ended up in locations unhealthier than their previous locations.[20] Families no longer had to rely exclusively on peddlers traveling through isolated areas to provide dry goods and food. And while it took very little to improve the poor housing conditions for renters and sharecroppers in north Alabama, the housing conditions did improve for some residents. Jess Panel's family ended up "much better situated than in their former location, as they have lived in nothing but a shack for the last 2 years."[21]

TVA also contributed to improved education in north Alabama. Education was thought to improve because children were now living, in many

Table 4. Homeownership in TVA reservoir areas, 1940 census

Counties	Owned (percent)	Percent change from 1930
Colbert	25.0	-8.0
Franklin	40.0	-3.4
Jackson	40.0	—
Lauderdale	38.5	-1.5
Lawrence	31.4	-1.2
Limestone	30.6	-0.6
Madison	26.8	-1.4
Marshall	40.2	-2.3
Morgan	35.9	-4.6

cases, closer to the schools they attended. Across the state, school attendance increased 2 percent, to 55 percent total attendance, among rural farm residents from 1930 to 1940.[22] The TVA noted that during the relocation process, "county school boards were particularly affected, since population changes naturally disrupt school attendance," but schools eventually were improved through the use of TVA money.[23] The TVA's library services became important parts of the community. Their mobile library units, what became known as bookmobiles, traveled up and down the river following the various construction projects.[24] Mary Utopia Rothrock, head of TVA libraries, called the impact of bookmobiles at the Wheeler and Pickwick construction sites "unbelievable. . . . Boy, they read . . . good things" out in the damtowns.[25] The influence this aspect of TVA operations had on the population's overall education "can scarcely be overestimated."[26]

TVA programs reached beyond the Tennessee Valley. Dam architects, particularly Roland Wank, were lauded for developing a clean design for future dams constructed across the country. Director of Land Planning and Housing Dr. Earle Draper once noted that before the TVA originated, dams were generally designed under the Greek or Gothic tradition, with embellishments that were unnecessary. Roland Wank and his team developed a new style consisting of "the smooth use of concrete without embellishment . . . which in a later review by the *Architectural Forum* was noted as one of the most important design contributions ever made to architecture in this country."[27]

The economic benefits were undeniable according to Paul Conner, who said, "cotton prices went up, corn prices went up, [and] things became available, like the electricity and the refrigerator."[28] The TVA did not intend or attempt to put an end to the sharecropper system, but they arguably contributed to the eventual reduced dependency on farming as a primary occupation. Though the economy did gradually recover from the Great Depression, perhaps in part to the TVA's influence, per capita income in Alabama had dropped from $331 per year to $281 per year in 1940. By 1941, Alabama's per capita income had jumped to $379 per year.[29] Paul also cited the cessation of floods and control of the river as important, not just for the overall quality of the land, but also because it meant areas would no longer become isolated due to floodwaters. Paul's cousins, Bobbie Curry and T. L. Conner, agreed that schools and land use improved. Paul and Dixie both agreed that the cash the TVA paid for Paul's father's land was important. Dixie commented, "I don't think you can discount what the cash money meant to Mr. Conner, because he'd never had that kinda money at one time." Paul remembered that on the forty acres his father eventually purchased, "we made some great crops after we moved out there."[30] Their house improved, too; they went from a four-room to a six-room house, which was a substantial upgrade. Some residents, like these families, moved beyond their fear and uncertainty to acceptance, making the best of a situation over which they had no control. Eugene Simonson felt that the "TVA is the best thing that ever happened to this part of the country" due to the economic benefits and infrastructure improvements, including electricity.[31]

As case files and interviews reveal, however, the benefits of relocation were bittersweet and often came with mixed feelings about the move and TVA in general. Substantial human costs resulted from the construction of Wheeler, Pickwick, and Guntersville Dams. It is impossible to quantify the turmoil experienced by the families whose lives were disrupted by forced removal. It's difficult to understand the cognitive dissonance a farmer must have felt when he was given a job (though temporary) for a government project that would potentially render his family homeless, with no promise of sustained employment or land to farm. Caseworkers responsible for ensuring that families relocated were asked to determine the families' attitudes toward the TVA before moving. What a complicated answer that must have been for the mother of five children, all of whom were malnourished, who hoped for a healthier and happier future but faced the prospect of leaving her home with no concrete resettlement plans. When the TVA "men" showed up on the front porch of an elderly man's ramshackle house on a humble farm, one he'd worked for decades, and told him they were buying his land for a nonnegotiable $50 per acre, he must have had mixed emotions. On one hand, the farmer would receive a paycheck unlike any he'd seen before, but he would be forced to leave the only land and lifestyle he knew.

The TVA's benefits did not always come immediately to farm families. By 1940, 26 percent of rural Alabama farm homes still had no running water, and 25 percent had no indoor toilet.[32] Paul Conner "still had to walk two miles to catch a school bus when [he] moved out there, and ride 20 miles before [he] got to school." Some relocated families would find improved housing but be left to deal with less fertile farmland. Or a family would find themselves with enough land to grow crops, but poor housing, or in an area away from their families and neighbors. The impacts of relocation were complicated, and it's difficult to generalize the varied experiences of the 2,500 families in north Alabama who had to relocate.

Residents' feelings about relocation were justifiably intense and often po-larized. As Bill Hardin summarized his take on what people thought about the TVA: "The ones that didn't go to work for TVA didn't have anything good to say about it. They were, uh, they were completely dissatisfied. . . . [Some] families hated the TVA and a lot of 'em thought it was the greatest thing since peanut butter!"[33] This enthusiasm came despite the amount of time it took to receive TVA power in some of the most outlying areas, like Hambrick Holler, where he and his people lived.

North Alabamians also lost more than just land and ancestral homes in the forced removal. For those whose livelihoods were negatively impact-ed—by reducing the amount of land they had to farm, removing them from the riverbanks where they could fish, or even taking away land that was used for growing corn for illegal whiskey—relocation resulted in the drop-ping of an old way of life and the need to quickly find a new way to make a living. Changing the depth of the water in Muscle Shoals resulted in the loss of the industry for which the town was named; the mussel-harvesting industry essentially disappeared after the TVA took control of the river.[34] In addition, archaeological expeditions undertaken during prescouting of the areas in which the dams would be constructed revealed Native American treasures and sacred burial mounds. These sites were studied extensively, the studies even framed as a positive result of the TVA. One newspaper article quoted a Tennessee archaeologist who noted that while the circum-stances necessitating the study were unfortunate, without the agency, "the prehistory of Tennessee might never have been known had it not been pos-sible to secure relief labor with which to excavate the buried villages which now lie many feet below the waters of the Chickamauga Lake."[35]

A Chattanooga Chamber of Commerce member reportedly once said of the TVA that "there is no question but that this agency has contribut-ed more to the economic development of the Tennessee Valley than any other single influence."[36] But, former Alabama Power chairman Thomas Martin wrote, "The future will have to appraise the social and economic value of TVA and its consequences."[37] Some feel that the changes credited to the TVA, including rural electricity and a better economy, would have happened with or without the TVA. Just how much credit is due the TVA for the South's modernization is open to debate. In her book on the TVA, Marguerite Owen wrote, "Some years ago, a thoughtful visitor asked the editor of a Valley newspaper if the changes might not have occurred without TVA. He reflected a moment before he made is oft-quoted reply: 'Well, they didn't.'"[38] Eugene Simonson said, "It is hard to imagine [the area without the TVA's influence]. Of course, WWII would've shaken up this area too, so I don't know how much of this quote 'progress' was due to WWII and how much was due to TVA, but both contributed, and when I say progress, it's not all progress, but change."[39] Some north Alabama residents recognized that the TVA years coincided with other important social changes that would have affected the Tennessee Valley regardless of whether the TVA purchased their lands. Newspapers frequently printed stories focusing on the correlation between societal improvements and the TVA. Whether or not there was a causal relationship is certainly debatable. But there is a question of whether the agency would have achieved its goals in north Alabama so quickly or so peacefully had it not been for the support of newspaper coverage and other media messages offering full support of the TVA.

The TVA was described repeatedly as a "grassroots" effort.[40] Rural residents facilitated the land-clearing efforts, families were displaced due to the purchase of massive amounts of land, and the TVA depended on rural residents to make the program successful by purchasing the cheap electricity made available to them after the dams started generating power. However, former TVA personnel director Gordon Clapp questioned calling the program a "grassroots" effort, given that the programs were imposed on Tennessee Valley residents rather than being a movement that started with rural families, the true definition of the term grassroots. The TVA was given permission to do whatever it deemed necessary for social, economic, and geographic improvements in the Tennessee Valley, and not all residents entirely supported its efforts. The TVA changed landscapes, lives, livelihoods, economies, and entire cities, with no guarantee of success. Thus, the TVA program really cannot be called a grassroots effort. It is more an example of an authoritative government agency that required the active participation of the residents it served to be successful.

Eugene Simonson, whose family moved from Arkansas to Guntersville so his father could work for the TVA, admitted that some in north Alabama still harbor ill will toward the TVA, saying "you can probably still find some old timers around here that'll tell you that TVA took the best of their land, and the best land is underwater here now." Simonson's former acquaintance "was all upset with TVA, and he mentioned that TVA took all the best farmland and it was all under water now, so it's an emotional kinda thing." The emotional turmoil of those who had to sacrifice their land is undeniable, and their sacrifices and hardships largely went unnoticed during the early TVA years. Addressing later TVA critics, particularly environmentalists who attacked the agency's excessive generation of power, Assistant Director of Information Paul Evans noted that many who criticized the TVA had never experienced a major economic depression on the scale of the Great Depression and, therefore, had no real right to attack the agency that helped pull the South of poverty. Many of them, he said, "had never known bread lines and unemployment. They had never known kerosene lamps. They had never known wash-tub bathrooms and a path. They weren't raised that way. . . . My feeling was that some . . . weren't even sure that these dams should have been built. This was like heresy to a person like myself."[41] Despite the hardships the TVA caused many families, they remained true to their ultimate goal of promoting the "well-being and security of the people of the area."[42]

Hindsight and time seem to have healed some of the wounds of the relocated. After the TVA forced his family from Hambrick Holler, Bill Hardin reflected on how his family's lives changed, saying, "I don't think it got any better. I don't think it got any worse. . . . We just moved from one farm to another," which was 2.5 miles away.[43] Beryl Tidwell felt that "for those people who had to move [life] might not have been better but for those that didn't have to, it was." When asked if her life improved because of the TVA program, Beryl Tidwell said with a laugh, "Well if it hadn't been for TVA, I never woulda [sic] met [my husband] if TVA hadn't bought the farm over there!" In Luther Tidwell, she found a kindred spirit and life partner; "we's both raised up about alike, poor, didn't have anything, [so we] worked hard and helped each other. I helped him in the field to haul beans, and he helped me out around the house."[44] The Tidwells eventually settled on a farm, the same land his father bought with the $9,999 the TVA had paid him for his original land. Their farm was productive for decades. Hazel Moore Thompson and Dixie Conner felt the TVA's intervention was similarly positive. Hazel Moore Thompson wondered "who I'da [sic] married if we'd stayed!"[45] Dixie Conner agreed that "if [my husband's family] hadn't had gone to Sand Mountain, none of this would've happen[ed], because he would've never known me."[46] Both women led lives as farm wives, working the land and raising families in north Alabama.

The scars of the TVA battle remain among those who endured physical and emotional upheaval, who felt that their lives were not improved by moving and that the prolonged hardships they faced tempered any benefits they eventually received. Family removal in north Alabama illustrates how it is possible for minority voices to be silenced when larger forces are working against them. In this case, a powerful government agency and a powerful media institution were able to keep a contingency of disadvantaged people quiet. Despite any long-term benefits the TVA may have provided, the voices of the relocated were not heard; they were not part of the TVA narrative largely constructed by professionals in the Information Office. This represents a failing of north Alabama newspapers to fulfill their public watchdog role, necessary for democracy to function effectively. Because newspapers were pro-TVA, relocated families had no way to speak out and no way to find assistance beyond what the TVA chose to offer.

The TVA made demands of Maxine Williamson Black a second time in the 1960s. She inherited some of their original farm her father had retained after the TVA bought his land, and eventually they learned of a forthcoming nuclear power plant: Browns Ferry in Athens, Alabama. She said the announcement to build the nuclear plant surprised her family. One day "somebody knocked on the door and it turned out to be a representative from TVA who said that they wanted to have the right of way" to relocate a road that would bisect the farm she owned. This time, she retained the services of a lawyer to investigate whether she could stop the TVA from cutting her beloved farm in half. In the end, she had no choice but to take the agency's offer and watch the TVA build a road right through the middle of her farm. For what they paid her, she said, "I didn't think it was [worth it] to cut the farm in two, because I love land so well. We tried [to stop it], we thought about it and talked about it and I went to talk to a lawyer and he said, 'Well, Maxine it's really nothin' you can do, because they can condemn it and take it for a dollar.' So I said, 'Well, if that's the case then I guess we'll just do like they did when they took the land down here for the water. We'll just take whatever they offer and look the other way and try to do the best you can.'"[47] Shortly thereafter, she started working at the newly constructed Browns Ferry Nuclear Power Plant. Mrs. Black's positive, "if you can't beat them, join them" attitude, much like her father's, enabled her to endure the changes the TVA brought to her dearly loved land.

Several aspects of relocation would have made for interesting news, some of which would have even readily fit within the existing pro-TVA stance. Instead, media messages placed in newspapers, magazine articles, newsreels, and radio programs reinforced the government's authority over rural residents. All the Information Office's efforts focused on helping the agency earn legitimacy and credibility among the public it would soon serve, despite the fact it was essentially a great social experiment. But north Alabama newspapers did not question the TVA. They did not examine what would happen if the TVA experiment failed or acknowledge the potential problem of what would happen if residents living along the river failed to move out of the way of the dams. They ignored concerns about the TVA's far-reaching authority to change the landscape and people's lives, those who had to sacrifice something of value to make room for TVA building projects were not acknowledged. In fact, newspapers went so far as to belittle those who spoke out against the agency. Newspapers directed readers' attention to the positive facets of TVA programs and failed to encourage critical thought or discussion. Public opinion left little room to doubt the improvements brought about in the region by the TVA.

Families forced off their land to make way for TVA projects faced considerable uncertainty, and neither the TVA nor the local newspapers offered much more to them than the vague promise that everything would eventually work out for their benefit. Neither institution had concrete evidence to support their claims that the TVA would be the institution that saved the South. Would they be able to make their new land as productive as their old land? Would they be able to afford the electricity the TVA planned to produce, even if it was cheap? How would their lives change? No one knew what would happen when the TVA's plans became reality, nor could they envision what the Tennessee Valley might look like when the TVA was finished with its construction projects. The TVA's regional planning efforts did not include making plans for relocated families beyond getting them to agree to move off their land. The TVA essentially left them to fend for themselves. Though the TVA coordinated with other relief agencies to help those who were in the most desperate need for help, assistance was discontinued once the families had left the reservoir areas. And, once the families left, newspapers continued to praise the TVA for completing projects on schedule.

The TVA's family removal procedures caused temporary hardships, struggles, and tough decisions for the people who lived on the land deemed necessary for dam construction. Despite the number of people directly impacted by family relocation—2,500 families in north Alabama alone—newspapers strategically omitted coverage of this aspect of TVA programs because it did not fit with the larger pro-TVA narrative, and because the Information Office was a powerhouse of publicity providing media content that emphasized the benefits to the region over the hardships relocated residents faced. Those who were forced to relocate are the forgotten heroes of the TVA project, people who were offered almost no public thanks and who were expected to give their land to the government without question. Still today, sacrifices like those made by relocated families are rarely acknowledged in media messages, yet those who give up their land so government and private-sector projects can proceed are contributing to the greater good, and they deserve to be praised. When a new runway at a major airport is approved, or a new section of an interstate highway is built—in fact, when any project that's characterized as an "improvement" is undertaken—we would all do well to consider the human costs.

Notes

Introduction

1. For a more complete summary of the history of the native people of Alabama, see Daniel Savage Gray, J. Barton Starr, and Linda Crockett Gray, *Alabama: A Place, A People, A Point of View* (Dubuque, IA: Kendall/Hunt Pub. Co., 1977).

2. Tennessee Valley Authority (TVA), *A History of Navigation on the Tennessee River: An Interpretation of the Economic Influence of This River System on the Tennessee Valley* (Washington, DC: TVA, 1937).

3. *Report on Economic Conditions of the South* (Knoxville, TN: National Emergency Council, 1938).

4. TVA, *A History of Navigation on the Tennessee.*

5. Hazel Moore Thompson, interview by Laura Beth Daws, June 16, 2012.

Chapter 1

1. Maxine Williamson Black, interview by Laura Beth Daws, June 17, 2012.

2. Family case record, "Cottingham, Austin," and "Cottingham, Lonnie," TVA Population Removal Records, Family Relocation Files 1934–1954, RG 142 Box 52, National Archives Southeast Region–Atlanta (National Archives–SERA).

3. James Sea Brown Jr., *Up before Daylight: Life Histories from the Alabama Writers' Project, 1938–1939* (Tuscaloosa: University of Alabama Press, 1982); James Agee and Walker Evans, *Let Us Now Praise Famous Men* (Boston: Houghton Mifflin, 1941); William Bradford Huie, *Mud on the Stars* (Tuscaloosa: University of Alabama Press, 1996); Wayne Flynt, *Poor but Proud: Alabama's Poor Whites* (Tuscaloosa: University of Alabama Press, 1989).

4. *Report on Economic Conditions of the South*, 8.

5. Brown Jr., *Up before Daylight*; Flynt, *Poor but Proud.*

6. Michael J. McDonald and John Muldowny, *TVA and the Dispossessed: The*

Resettlement of Population in the Norris Dam Area (Knoxville: University of Tennessee Press, 1981).

7. Flynt, *Poor but Proud*, 60.

8. TVA, Research Section, Social and Economic Division. *Preliminary and Confidential Report: Families of the Wheeler Reservoir Area.* (Knoxville, TN: TVA), September 12, 1935, 4–5.

9. Wilson Whitman, *God's Valley* (New York: Viking Press, 1939).

10. Charles Kenneth Roberts, *The Farm Security Administration: Rural Rehabilitation in the South* (Knoxville: University of Tennessee Press, 2015).

11. Herman C. Nixon, *Forty Acres and Steel Mules*, 26, as found in Flynt, *Poor but Proud*, 78.

12. US Census Bureau, 1930 census.

13. *Report on Economic Conditions of the South.*

14. Nancy Cabaniss Parker, interview by Laura Beth Daws, July 16, 2012.

15. Paul Conner, interview by Laura Beth Daws and Susan Brinson, August 6, 2012.

16. Nancy Grant, *TVA and Black Americans: Planning for the Status Quo* (Philadelphia: Temple University, 1990), 15.

17. US Census Bureau, 1940 census.

18. US Census Bureau, "Farms—Number, Acreage and Value, by Color of Operator for North and West, and by Color and Tenure of Operators, for South, by States: 1930 and 1935," Table 616, http://www2.census.gov/prod2/statcomp/documents/1940-07.pdf (accessed July 3, 2014).

19. *Report on Economic Conditions of the South.*

20. Bill Hardin, interview by Laura Beth Daws, July 15, 2012.

21. Maxine Williamson Black, interview by Laura Beth Daws, June 17, 2012.

22. US Census Bureau, 1930 census, *Population*, vol. 3, part 1, Tables 30 and 38.

23. TVA, *Preliminary and Confidential Report–The Guntersville Area and the Proposed Coles Bend Bar Dam* (Knoxville, TN: TVA), 22.

24. William Warren Rogers, Robert David Ward, Leah Rawls Atkins, and Wayne Flynt, *Alabama: The History of a Deep South State* (Tuscaloosa: University of Alabama Press, 1994).

25. TVA, *Preliminary and Confidential Report–Guntersville.* Colbert, Jackson, Lauderdale, Limestone, Madison, Marshall, and Morgan Counties were included in this figure.

26. Flynt, *Poor but Proud*, 295.

27. *Report on Economic Conditions of the South*, 21.

28. W. M. Adamson, *Income in Counties of Alabama, 1929–1935* (Tuscaloosa: Bureau of Business Research, School of Commerce and Business Administration, University of Alabama, 1939), 6.

29. Adamson, *Income in Counties of Alabama.* Averages for Wheeler counties include Colbert, Lauderdale, Lawrence, Limestone, Madison, and Morgan. Averages for Guntersville include Marshall and Jackson.

30. US Census Bureau, 1930 census, "Hours and Earnings," Table 374.

31. *Report on Economic Conditions of the South*, 22.

32. Family case record, "Simms, Janet," TVA Population Removal Records, Family Relocation Files 1934–1954, RG 142 Box 82, National Archives–SERA.

33. Rogers et al., *Alabama: The History of a Deep South State*.

34. Family case record, "Torance, James." TVA Population Removal Records, Family Relocation Files 1934–1954, RG 142 Box 58, National Archives–SERA.

35. "Unemployment Relief Census, October 1933," *Monthly Labor Review* 39 (July 1934): 31.

36. Family case record, "Preston, Emmett," TVA Population Removal Records, Family Relocation Files 1934–1954, RG 142 Box 68, National Archives–SERA.

37. US Census Bureau, 1930 census.

38. US Census Bureau, 1930 census, vol. 6, *Population & Families*, Table 18.

39. Family case record, "Moore, Sam," TVA Population Removal Records, Family Relocation Files 1934–1954, RG 142 Box 81, National Archives–SERA.

40. Family case record, "Cottingham, Austin," and "Cottingham, Lonnie."

41. Family case record, "Stone, Sam," TVA Population Removal Records, Family Relocation Files 1934–1954, RG 142 Box 82, National Archives–SERA.

42. *Report on Economic Conditions of the South*, 34.

43. Family case record, "Kight, John," TVA Population Removal Records, Family Relocation Files 1934–1954, RG 142 Box 80, National Archives–SERA.

44. Family case record, "Green, Lessie," TVA Population Removal Records, Family Relocation Files 1934–1954, RG 142 Box 53, National Archives–SERA.

45. Family case record, "Holcomb, Bill," TVA Population Removal Records, Family Relocation Files 1934–1954, RG 142 Box 80, National Archives–SERA.

46. Family case record, "Etters, Robert," TVA Population Removal Records, Family Relocation Files 1934–1954, RG 142 Box 52, National Archives–SERA.

47. Family case record, "Whitehead, Arie," TVA Population Removal Records, Family Relocation Files 1934–1954, RG 142 Box 82, National Archives–SERA.

48. Family case record, "Russell, Willie," TVA Population Removal Records, Family Relocation Files 1934–1954, RG 142 Box 68, National Archives–SERA.

49. Family case record, "Anderson, Jim Don," TVA Population Removal Records, Family Relocation Files 1934–1954, RG 142 Box 78, National Archives–SERA.

50. Family case record, "Andrews, Elijah (Box 50)"; "Cloud, Charlie (Box 52)"; "Frazier, Will (Box 52)"; and "Ledbetter, Virgil (Box 55)," TVA Population Removal Records, Family Relocation Files 1934–1954, RG 142, National Archives–SERA.

51. Family case record, "Green, Thomas," TVA Population Removal Records, Family Relocation Files 1934–1954, RG 142 Box 53, National Archives–SERA.

52. Family case record, "Rice, Dave, Jr." TVA Population Removal Records, Family Relocation Files 1934–1954, RG 142 Box 56, National Archives–SERA.

53. Family case record, "Torance, James."

54. *Report on Economic Conditions of the South.*

55. *Report on Economic Conditions of the South*, 29.

56. *Report on Economic Conditions of the South*, 29.

57. Lawrence L. Durisch, *Preliminary and Confidential Report: Families of the Wheeler Reservoir Area* (Knoxville, TN: TVA), 11.

58. *Report on Economic Conditions of the South*, 34.

59. TVA, *Preliminary Report–Guntersville*, 20.

60. Family case record, "Tidwell, Clarence," TVA Population Removal Records, Family Relocation Files 1934–1954, RG 142 Box 82, National Archives–SERA.

61. Family case record, "Durham, George," TVA Population Removal Records, Family Relocation Files 1934–1954, RG 142 Box 52, National Archives–SERA.

62. Bessie Barker, "A Study of the Living Conditions of One Hundred Families in Limestone County, Alabama on Various Economic Levels" (master's thesis, Alabama Polytechnic Institute, 1934).

63. Family case record, "Torance, James."

64. Family case record, "Ledbetter, Virgil."

65. Family case record, "Redmon, Barney," TVA Population Removal Records, Family Relocation Files 1934–1954, RG 142 Box 81, National Archives–SERA.

66. Bill Hardin, interview by Laura Beth Daws.

67. Beryl Tidwell, interview by Laura Beth Daws, July 16, 2012.

68. Family case record, "Risner, Dock (Box 56)"; "Torance, James (Box 58)"; and "Baker, Jack," TV (Box 51), Population Removal Records, Family Relocation Files 1934–1954, RG 142, National Archives–SERA.

69. T. L. Conner, interview by Laura Beth Daws, July 19, 2012.

70. Family case record, "Anderson, Lucy," TVA Population Removal Records, Family Relocation Files 1934–1954, RG 142 Box 58, National Archives–SERA.

71. Family case record, "Ball, T. D.," TVA Population Removal Records, Family Relocation Files 1934–1954, RG142 Box 78, National Archives–SERA.

72. Flynt, *Poor but Proud*, 164.

73. Family case record, "Hardin, Emma Kate," TVA Population Removal Records, Family Relocation Files 1934–1954, RG 142 Box 80, National Archives–SERA.

74. TVA, *Agricultural-Industrial Survey of Marshall County Alabama*, ed. Industry Division of TVA (Knoxville, TN: TVA,1935), 4.

75. *Report on Economic Conditions of the South*, 22.

76. *Report on Economic Conditions of the South*, 22.

77. US Census Bureau, "Statistical Summary of Education," 1940 census.

78. US Census Bureau, 1940 census.

79. TVA, *Agricultural–Industrial Survey of Marshall County*, 4.

80. TVA, *Preliminary Report–Guntersville*, 21.

81. For a more complete history of relief efforts in north Alabama, see Wayne

Flynt, *Alabama in the 20th Century* (Tuscaloosa: University of Alabama Press, 2006), or Rogers et al., *Alabama: The History of the Deep South State.*

82. TVA, *Preliminary Report–Guntersville*, 4.

83. TVA, *Preliminary Report–Guntersville.*

84. Gray et al., *Alabama*, 287.

Chapter 2

1. Donald H. Grubbs, *Cry from the Cotton: The Southern Tenant Farmers' Union and the New Deal* (Chapel Hill: University of North Carolina Press, 1971); Arthur M. Schlesinger, *The Coming of the New Deal* (Boston: Houghton Mifflin, 1959).

2. Roger Biles, *The South and the New Deal* (Lexington: University Press of Kentucky, 1994), 35. See also Pete Daniel, "The New Deal, Southern Agriculture, and Economic Change," in *The New Deal and the South*, ed. James C. Cobb and Michael V. Namorato (Jackson: University Press of Mississippi, 1984), 37–61.

3. Daniel, "The New Deal, Southern Agriculture, and Economic Change."; Frank Freidel, "The South and the New Deal," in *The New Deal and the South*, ed. James C. Cobb and Michael V. Namorato (Jackson: University Press of Mississippi, 1984), 17–36; Paul E. Mertz, *New Deal Policy and Southern Rural Poverty* (Baton Rouge: Louisiana State University Press, 1978).

4. Roberts, *The Farm Security Administration: Rural Rehabilitation in the South.*

5. David E. Conrad, *The Forgotten Farmers: The Story of Sharecroppers in the New Deal* (Urbana: University of Illinois Press, 1965); Paul E. Mertz, *New Deal Policy and Southern Rural Poverty.*

6. David E. Conrad, *The Forgotten Farmers: The Story of Sharecroppers in the New Deal*, 35. See also, Warren C. Whatley, "Labor for the Picking: The New Deal in the South," *Journal of Economic History* 43, no. 4 (1983): 905–929.

7. Daniel, "The New Deal, Southern Agriculture, and Economic Change," 52.

8. Anthony Badger, *New Deal/New South* (Fayetteville: University of Arkansas Press, 2007), 32.

9. Badger, *New Deal/New South*, 38. See also Biles, *The South and the New Deal*, 35. See also James C. Cobb and Michael V. Namorato, "Introduction," in *The New Deal and the South*, ed. James C. Cobb and Michael V. Namorato (Jackson: University Press of Mississippi, 1984), 3–15.

10. See, for example, Jacqueline Jones, "Federal Power, Southern Power: 1860–1940," *Journal of American History* 87, no. 4 (2001): 1392–1396.

11. Biles, *The South and the New Deal*, 58.

12. Ronald E. Seavoy, *The American Peasantry: Southern Agricultural Labor and Its Legacy, 1850–1995* (Westport, CT: Greenwood Press, 1998); Grubbs, *Cry from the Cotton.*

13. Paul E. Mertz, *New Deal Policy and Southern Rural Poverty*, 255–256.

14. Daniel, "The New Deal, Southern Agriculture, and Economic Change," 60.

15. Badger, *New Deal/New South: An Anthony Badger Reader*; Dewey W. Grantham, "Regional Claims and National Purposes: The South and the New Deal," *Atlanta History: A Journal of Georgia and the South* 38, no. 3 (1994): 5–17; Gavin Wright, "The New Deal and the Modernization of the South," *Federal History*, 2010, http://shfg.org/shfg/wp-content/uploads/2011/01/5-Wright-design5-_Layout-1.pdf (accessed November 26, 2017).

16. Leah Rawls Atkins, *Developed for the Service of Alabama: The Centennial History of the Alabama Power Company, 1906–2006* (Birmingham: Alabama Power Company, 2006), 62.

17. For a more thorough discussion of the history of Wilson Dam, see Preston J. Hubbard, *Origins of the TVA: The Muscle Shoals Controversy, 1920–1932* (Nashville, TN: Vanderbilt University Press, 1961).

18. Rogers et al., *Alabama: The History of a Deep South State.*

19. Atkins, *History of the Alabama Power Company.*

20. Hubbard, *Origins of the TVA.*

21. Hubbard, *Origins of the TVA.*

22. Atkins, *History of the Alabama Power Company*; George Norris, *Fighting Liberal: The Autobiography of George W. Norris* (New York: Collier Books, 1961).

23. Philip Selznick, *TVA and the Grass Roots* (Berkeley: University of California Press, 1984), 5.

24. Selznick, *TVA and the Grass Roots.*

25. William U. Chandler, *The Myth of TVA: Conservation and Development in the Tennessee Valley, 1933–1983* (Cambridge, MA: Ballinger Publishing Company, 1984), 38.

26. Jean Edward Smith, *FDR* (New York: Random House, 2008).

27. HR 5081, 73rd Cong., 1st sess., May 18, 1933.

28. James A. Hagerty, "Alabamans Cheer Him," *New York Times*, January 22, 1933.

29. TVA Board of Directors, *Report to Congress on the Unified Development of Tennessee*, 1936, 24.

30. Marguerite Owen, *The Tennessee Valley Authority* (New York: Praeger), 75–76.

31. TVA Reservoir Property Management Population Removal Records, Administrative Files, Cooperative Relations between TVA and Outside Agencies, RG 142 Box 6, National Archives–SERA.

32. Smith, *FDR*, 325.

33. Smith, *FDR.*

34. R. L. Duffus, *The Valley and Its People* (New York: Knopf, 1946), 57.

35. Aaron D. Purcell, *Arthur Morgan: A Progressive Vision for American Reform* (Knoxville: University of Tennessee Press, 2014), 136.

36. Arthur E. Morgan, *Making of the TVA* (Buffalo, NY: Prometheus Books, 1974), 55.

37. Chandler, *Myth of TVA*, 32.

38. Morgan, *Making of the TVA*, 55.

39. David E. Lilienthal, interviews by Charles W. Crawford, February 6 and 7, 1970.

40. Lilienthal, interview by Charles W. Crawford.

41. Purcell, *Arthur Morgan*, 148.

42. R. L. Duffus, "The Storm Center of the TVA's Roaring Dams," *New York Times*, April 3, 1938, 119.

43. Mary Utopia Rothrock, interview by Charles W. Crawford, January 16, 1970.

44. Lilienthal, interview by Charles W. Crawford.

45. Earle Sumner Draper, interview by Charles W. Crawford, December 30, 1969.

46. Charles McCarthy, interview by Charles W. Crawford, October 30, 1969.

47. Charles Krutch, interview by Charles W. Crawford, November 10, 1969.

48. J. Dudley Dawson and Arthur E. Morgan, interview by Charles W. Crawford, June 20, 1969.

49. Edward Falck, interview by Charles W. Crawford, September 25, 1970.

50. Lilienthal, interview by Charles W. Crawford.

51. Grant, *TVA and Black Americans*, 32.

52. Charles Krutch, interview by Charles W. Crawford.

53. Smith, *FDR*.

54. Hagerty, "Alabamans Cheer Him."

55. For more historical context, see William E. Leuchtenburg, *Franklin D. Roosevelt and the New Deal, 1932–1940* (New York: Harper & Row, 1963).

56. TVA Board of Directors, *Report to Congress*, 32.

57. TVA, Office of Engineering, Design, and Construction, Engineering Project Reports, RG 142 Box 32, File 468, National Archives–SERA.

58. Owen, *The Tennessee Valley Authority*.

59. David E. Lilienthal, *Democracy on the March* (New York: Harper & Row, 1953).

60. Selznick, *TVA and the Grassroots*.

61. Smith, *FDR*, 325.

62. "Muscle Shoals, 1933."

63. Aaron D. Purcell, *White Collar Radicals* (Knoxville: University of Tennessee Press, 2009), 15.

64. For a complete explanation of the parallels between the TVA and its communist roots, see Purcell, *White Collar Radicals*.

65. Earle Sumner Draper, interview by Charles W. Crawford.

66. Leuchtenburg, *Franklin D. Roosevelt and the New Deal, 1932–1940*.

67. Fannon Beauchamp, interview by Charles W. Crawford, June 16, 1970.

68. Fannon Beauchamp, interview by Charles W. Crawford.

69. Whitman, *God's Valley*, 215.

70. "To Hold Cooking School," *Limestone (AL) Democrat*, August 15, 1935, 5.

71. Gregory B. Field, "'Electricity for All': The Electric Home and Farm Authority and the Politics of Mass Consumption, 1932–1935." *The Business History Review* 64, no. 1 (1990): 32–60.

72. Field, "Electricity for All," 32–60.

73. Smith, *FDR*, 324; quote from William Starr Myers, ed., *The State Papers of Herbert Hoover* (New York: Doubleday, Doran 1934), 526–527.

74. Fannon Beauchamp, interview by Charles W. Crawford.

75. R. L. Duffus, *The Valley and Its People*, 64.

76. For more information about Reddy Kilowatt's origins, see Atkins, *Developed for the Service of Alabama: The Centennial History of the Alabama Power Company, 1906–2006.*

77. "Alabama Power," *Guntersville (AL) Advertiser & Democrat*, March 30, 1939.

78. "Rural Electrification in Alabama," *Albertville (AL) Herald*, October 29, 1936, 5.

79. "Rural Electrification in Alabama."

80. Smith, *FDR*, 325.

81. "Supreme Court Upholds TVA," *Albertville Herald*, February 20, 1936, 1.

82. *New York Times*, February 18, 1936, 1.

Chapter 3

1. TVA Population Removal Records, Family Relocation Files, 1934–1954, RG 142, National Archives–SERA.

2. Family case record, "Filmore, Isaac," TVA Population Removal Records, Family Relocation Files 1934–1954, RG 142 Box 79, National Archives–SERA.

3. Family case record, "Filmore, Clyde," TVA Population Removal Records, Family Relocation Files 1934–1954, RG 142 Box 79, National Archives–SERA.

4. TVA Office of the General Manager, Information Office Correspondence Files, 1933–1946, "TVA Housing Development at Pickwick Dam," RG 142 Box 23, National Archives–SERA.

5. Family case record, "Ward, Henry" TVA Population Removal Records, Family Relocation Files 1934–1954, RG 142 Box 70, National Archives–SERA.

6. This "protective belt" policy, however, was not in place officially until the beginning of the Guntersville Dam project (see Selznick, *TVA and the Grass Roots*).

7. TVA Reservoir Property Management Department, Population Readjustment Division, *Family Case Records Population Readjustment*, 2.

8. Selznick, *TVA and the Grass Roots*.

9. Earle Sumner Draper, interview by Charles Crawford.

10. Selznick, *TVA and the Grass Roots*, 100.

11. Earle Sumner Draper, interview by Charles Crawford.

12. TVA, *Technical Review of the Wheeler Project*, Technical monograph no. 38 (Knoxville, TN: TVA, 1938).

13. TVA, *The Guntersville Project: A Comprehensive Report on the Planning, Design, Construction and Initial Operations of the Guntersville Project*, Technical report no. 4 (Knoxville, TN: TVA), 1941; *Annual Report of the Tennessee Valley Authority for the Fiscal Year ended June 30 1935* (New York: Arno Press, 1969).

14. Earle Sumner Draper, interview by Charles Crawford.

15. Family case record, "Bottomley, Will," TVA Population Removal Records, Family Relocation Files 1934–1954, RG142 Box 51, National Archives–SERA.

16. Carl Kitchens, "Use of Eminent Domain in Land Assembly: The Case of the Tennessee Valley Authority," *Public Choice* 160, no. 3/4 (2014): 455–466.

17. *Tennessee Valley Authority Land Acquisitions Division Appraisal Section Manual*–Confidential, April 1936, 6; Office of Engineering, Design, and Construction, Engineering Project Histories and Reports, RG 142 Box 26, National Archives–SERA.

18. *Tennessee Valley Authority Land Acquisitions Division Appraisal Section Manual*–Confidential, 6.

19. Matthew Downs, *Transforming the South, Federal Development in the Tennessee Valley, 1915–1960* (Baton Rouge: Louisiana State University Press, 2014).

20. TVA Land Acquisition Division, *Instructions to Land Buyers*, 1936.

21. TVA Land Acquisition Division, *Instructions to Land Buyers*.

22. TVA, *Technical Review of the Wheeler Project*; TVA, *The Guntersville Project*; *Annual Report of the Tennessee Valley Authority for the Fiscal Year ended June 30 1935*, 1969.

23. Owen, *The Tennessee Valley Authority*, 67.

24. Owen, *The Tennessee Valley Authority*.

25. Charles McCarthy, interview by Charles Crawford.

26. Owen, *The Tennessee Valley Authority*.

27. Owen, *The Tennessee Valley Authority*, 67.

28. R. L. Duffus, *The Valley and Its People: A Portrait of TVA*, 59.

29. TVA, *Technical Review of the Wheeler Project*, 229; TVA, *The Guntersville Project*, 253.

30. Family case record, "Cole, Buddy," TVA Population Removal Records, Family Relocation Files 1934–1954, RG 142 Box 52, National Archives–SERA.

31. Family case record, "Cole, Buddy."

32. Luther Tidwell, interview by Laura Beth Daws, July 16, 2012.

33. McDonald and Muldowny, *TVA and the Dispossessed*.

34. "Big Tennessee Dam Will Bury Small Village and Several Cemeteries," *Guntersville Advertiser & Democrat*, May 3, 1933, 4.

35. TVA Reservoir Property Management Department, Population Readjustment Division, *Family Case Records Population Readjustment*, 2.

36. J. S. Beauchamp, Head Property Officer, Materials Division, Knoxville, Tennessee, letter to A. L. Snell, Family Removal Section, Decatur, Alabama, May 25, 1936, in "Neville, George," TVA Population Removal Records, Family Relocation Files 1934–1954, RG 142 Box 56, National Archives–SERA.

37. TVA, *Population Readjustment, Guntersville Area* (Knoxville, TN: TVA, June 1940), 1.

38. TVA, *Wheeler Project–Final Report (First Draft), Report on Family Removal Activities* (Knoxville, TN: TVA, 1936) 4, in TVA Population Removal Records, Family Relocation Files 1934–1954, RG142 Box 59, National Archives–SERA.

39. TVA, *Wheeler Project–Final Report (First Draft)*, 2.

40. TVA, *Wheeler Project–Final Report (First Draft)*, 3.

41. Family case record, "Whittaker, L. B.," TVA Population Removal Records, Family Relocation Files 1934–1954, RG142 Box 82, National Archives–SERA.

42. TVA, *Activities of the Reservoir Family Removal Section, Wheeler Reservoir Area*, 2, in TVA Population Removal Records, Family Relocation Files 1934–1954, RG142 Box 59, National Archives—SERA.

43. TVA, *Activities of the Reservoir Family Removal Section, Wheeler Reservoir Area*, 2.

44. TVA, *Activities of the Reservoir Family Removal Section, Wheeler Reservoir Area–First Draft*, 1.

45. All other caseworkers earned $1,800 for their work at Wheeler and anywhere from $1,800 to $2,600 in Guntersville.

46. TVA, *Activities of the Reservoir Family Removal Section, Wheeler Reservoir Area*, 6.

47. TVA, *Wheeler Project–Final Report, Report on Family Removal Activities*.

48. Family case record, "Craft, James," TVA Population Removal Records, Family Relocation Files 1934–1954, RG142 Box 52, National Archives–SERA.

49. Family case record, "Gilliam, James," TVA Population Removal Records, Family Relocation Files 1934–1954, RG142 Box 79, National Archives–SERA.

50. TVA, *Activities of the Reservoir Family Removal Section, Wheeler Reservoir Area–First Draft*, 1.

51. TVA, *Activities of the Reservoir Family Removal Section, Wheeler Reservoir Area–First Draft*, 1, Table 4.

52. TVA, *Activities of the Reservoir Family Removal Section, Wheeler Reservoir Area–First Draft*, 14.

53. Family case record, "Rogers, Russell," TVA Population Removal Records, Family Relocation Files 1934–1954, RG 142 Box 81, National Archives–SERA.

54. Bobbie Curry, interview by Laura Beth Daws, July 19, 2012.

55. Paul Conner, interview by Laura Beth Daws and Susan Brinson.

56. Paul Conner, interview by Laura Beth Daws and Susan Brinson.

57. Hazel Moore Thompson, interview by Laura Beth Daws.

58. Luther Tidwell, interview by Laura Beth Daws.

59. A. L. Snell, to Ed Sandlin, in Family case record, "Sandlin, Ed," TVA Population Removal Records, Family Relocation Files 1934–1954, RG142 Box 81.

60. Family case record, "Day, Roy," TVA Population Removal Records, Family Relocation Files 1934–1954, RG 142 Box 52, National Archives–SERA.

61. Family case record, "Terry, Fred," TVA Population Removal Records, Family Relocation Files 1934–1954, RG 142 Box 58, National Archives–SERA.

62. T. L. Conner, interview by Laura Beth Daws.

63. Luther Tidwell, interview by Laura Beth Daws.

64. Bill Hardin, interview by Laura Beth Daws.

65. Maxine Williamson Black, interview by Laura Beth Daws, June 7, 2012.

66. George Hodge, interview by Laura Beth Daws, July 16, 2012.

67. Bill Hardin, interview by Laura Beth Daws.

68. Family case record, "Cagle, W. P.," TVA Population Removal Records, Family Relocation Files 1934–1954, RG 142 Box 78, National Archives–SERA.

69. Family case record, "Floyd, J. W.," TVA Population Removal Records, Family Relocation Files 1934–1954, RG 142 Box 79, National Archives–SERA.

70. Family case record, "Ivy, Peter," TVA Population Removal Records, Family Relocation Files 1934–1954, RG 142 Box 54, National Archives–SERA.

71. Family case record, "Neville, George," TVA Population Removal Records, Family Relocation Files 1934–1954, RG 142 Box 56, National Archives–SERA.

72. "What It Means to Limestone," *Guntersville Advertiser & Democrat*, July 11, 1934, 2.

73. "Limestone County Sold Land to TVA for $4,355," *Limestone Democrat*, January 16, 1936.

74. "Activity in Real Estate," *Limestone Democrat*, August 30, 1934, 4.

75. "Big Cat Caught by Sheriff Wellden—Much Mash and Liquor Destroyed," *Alabama Courier* (Athens, AL), March 14, 1935, 1.

76. "Big Cat Caught by Sheriff Wellden—Much Mash and Liquor Destroyed."

77. "Col. Bibb Graves Visits Limestone," *Limestone Democrat*, October 11, 1934, 1.

78. "Taxation Figures Remain at Normal: TVA Land Acquisition Not Harmful," *Limestone Democrat*, April 23, 1936, 1.

79. "386,525.30 Total Paid for Local Tracts by TVA," *Huntsville Times*, February 22, 1936, 1.

80. "What It Means to Limestone."

81. "Three Executives in Charge of Uncle Sam's Development," *Decatur (AL) Daily*, March 14, 1934, 8.

82. "Reassuring," *Alabama Courier*, July 5, 1934, 4.

83. Untitled, *Alabama Courier*, July 19, 1934, 4.

84. "Where Will 'They' Locate the Dam," *Guntersville Advertiser & Democrat*, September 11, 1935, 1.

85. *Population Readjustment in the Tennessee Valley*, 6; TVA Office of the General Manager, Information Office Correspondence Files, 1933–1946, "TVA Housing Development at Pickwick Dam," RG 142 Box 23, National Archives–SERA.

Chapter 4

1. Charles Krutch, interview by Charles Crawford.

2. Charles Krutch, interview by Charles Crawford.

3. Charles Krutch, interview by Charles Crawford.

4. TVA, "Review of the Program and Policy of the Information Office," May 1940, in TVA Office of the General Manager, Information Office, Correspondence Files, 1933–June 1944, RG 142 Box 6, National Archives–SERA.

5. Charles Krutch, interview by Charles Crawford.

6. Joseph Swidler, interviews by Charles Crawford, October 28, 1969, and December 8, 1969.

7. Walter N. Polakov and Theodor Swanson, "Sample of an Outline for Organization of the Section 10," August 1, 1933, in TVA Office of the General Manager, Information Office Correspondence Files 1933–1946, RG 142 Box 7, National Archives–SERA.

8. Thomas K. McCraw, *TVA and the Power Fight, 1933–1939* (Philadelphia: J. B. Lippincott, 1971), 148.

9. TVA, "Review of the Program and Policy of the Information Office," 30.

10. TVA Office of the General Manager, Information Office Correspondence Files, 1933–June 1944, RG 142 Box 7, page 7, National Archives–SERA.

11. W. L. Sturdevant to Floyd Reeves, August 18, 1933. In TVA Office of the General Manager, Information Office Correspondence Files 1933–1946, RG 142 Box 7, National Archives–SERA.

12. McCraw, *TVA and the Power Fight*, 146.

13. Walter N. Polakov and Theodor Swanson, "Sample of an Outline for Organization of the Section 10," August 1, 1933, in TVA Office of the General Manager, Information Office Correspondence Files 1933–1946, RG 142 Box 7, National Archives–SERA.

14. W. L. Sturdevant, letter to David E. Lilienthal, September 3, 1935, in TVA Office of the General Manager, Information Office Correspondence Files, 1933–June 1944, National Archives–SERA.

15. Charles Krutch, interview by Charles Crawford.

16. TVA Office of the General Manager, Information Office Correspondence Files 1933–June 1944, RG142 Box 6, page 1, National Archives–SERA.

17. TVA Office of the General Manager, Information Office Correspondence Files 1933–June 1944, RG142 Box 6, National Archives-SERA.

18. Joseph Swidler, interviews by Charles Crawford.

19. David E. Lilienthal, letter to William Goodman, November 10, 1934, in TVA Office of the General Manager, Information Office Correspondence Files 1933–June 1944, RG142 Box 1, National Archives–SERA.

20. TVA Office of the General Manager, Information Office Correspondence Files 1933–June 1944, RG142 Box 6, page 11, National Archives–SERA.

21. Mary Utopia Rothrock, interview by Charles Crawford.

22. Edward Falck, interview by Charles Crawford.

23. Edward Falck, interview by Charles Crawford.

24. McCraw, *TVA and the Power Fight*, 148.

25. McCraw, *TVA and the Power Fight*, 146.

26. David E. Lilienthal, "Talk to TVA Employees" (speech delivered at Chickamauga Dam, Chattanooga, TN, July 5, 1937), in TVA Office Engineering, Design, and Construction, Engineering Project Histories and Reports, RG142 Box 32, National Archives–SERA.

27. TVA, Office of the General Manager, Information Office, Correspondence Files, 1933–June 1944, in RG 142 Box 7, page 11, National Archives–SERA.

28. TVA, Office of the General Manager, Information Office, Correspondence Files, 1933–June 1944, in RG 142 Box 7, National Archives–SERA.

29. McCraw, *TVA and the Power Fight*, 148.

30. Forrest Allen, letter to J. B. Blandford Jr., re: Employee Publications, August 16, 1935, in TVA Office of the General Manager, Information Office Correspondence Files 1933–June 1944, RG142 Box 11, National Archives–SERA.

31. J. V. Ankeney to J. B. Blandford Jr., re: Employee Publications, August 21, 1935, in TVA Office of the General Manager, Information Office Correspondence Files 1933–June 1944, RG142 Box 11, National Archives–SERA.

32. J. V. Ankeney to J. B. Blandford Jr., Re: Employee Publications, August 21, 1935.

33. Forrest Allen to J. B. Blandford Jr., letter re: Employee Publications, August 16, 1935.

34. "Information Office," booklet, September 7, 1937, in TVA Office of the General Manager, Information Office Correspondence Files 1933–June 1944, RG142 Box 7, page 8, National Archives–SERA.

35. "Review of the Program and Policy of the Information Office," 9.

36. William I. Nichols, "Teaching Grandmother How to Spin: TVA and the Private Utilities," *Harper's Magazine* (July 1936): 113–119.

37. Family case record, "Mefford, Dewey," TVA Population Removal Records, Family Relocation Files 1934–1954, RG 142 Box 81, National Archives–SERA.

38. Family case record, "Risner, Dock."

39. TVA Office of the General Manager, Information Office Correspondence Files, 1933–June 1944, RG 142 Box 6, page 9, National Archives–SERA.

40. Edward Falck, interview by Charles Crawford.

41. David E. Lilienthal, letter to Tennessee Valley residents, March 3, 1936, in TVA, Office of the General Manager, Records of the Board of Directors, David E. Lilienthal General Correspondence 1933–1946, RG142 Box 1, National Archives–SERA.

42. Public relations letters, in TVA, Office of the General Manager, Records of the Board of Directors, David E. Lilienthal General Correspondence 1933–1946, RG142 Box 1, National Archives–SERA.

43. David Lilienthal, interview by Charles Crawford.

44. David Lilienthal, interview by Charles Crawford.

45. David E. Lilienthal, "The Future of Industry in the Tennessee Valley Region," (speech delivered at the Tennessee Valley Institution of the University of

Chattanooga, Chattanooga, TN, April 21, 1934), in TVA, Office of Engineering, Design, and Construction, Engineering Project Histories and Reports, RG 142 Box 23, National Archives–SERA.

46. David E. Lilienthal, (speech delivered to the Cleveland, Ohio, Advertising Club, June 3, 1936), in TVA, Office of Engineering, Design, and Construction, Engineering Project Histories and Reports, RG 142 Box 23, National Archives–SERA.

47. David E. Lilienthal, "Figures Show TVA's Electricity Plan Works" (speech delivered to the American Society of Mechanical Engineers, Norris, TN, June 22, 1935), in TVA, Office of Engineering, Design, and Construction, Engineering Project Histories and Reports, RG 142 Box 32, National Archives–SERA.

48. David E. Lilienthal, "The Tennessee Valley Authority and Farm Electrification" (speech delivered to the National Convention of the American Farm Bureau Federation, Nashville, TN, December 12, 1934), in TVA, Office of Engineering, Design, and Construction, Engineering Project Histories and Reports, RG 142 Box 33, National Archives–SERA.

49. David E. Lilienthal, "Birmingham and the TVA Industrial Development Program" (speech delivered to the Birmingham Rotary Club, Birmingham, AL, October 31, 1934), in TVA, Office of Engineering, Design, and Construction, Engineering Project Histories and Reports, RG 142 Box 23, National Archives–SERA.

50. David E. Lilienthal, "Some Observations on the TVA" (speech delivered to TVA employees at First Methodist Church, Knoxville, TN, June 12, 1936), in TVA, Office of Engineering, Design, and Construction, Engineering Project Histories and Reports, RG 142 Box 23, National Archives–SERA.

51. Purcell, *Arthur Morgan*, 159.

52. "TVA Appreciation Jubilee July Fourth," *Limestone Democrat*, June 13, 1935, 5; "TVA Dinner Will Be Held Friday," *Limestone Democrat*, December 13, 1934, 1; "TVA Appreciation Jubilee at Muscle Shoals July Fourth," *Alabama Courier* June 13, 1935, 1; "Program on Use of Electricity Planned," *Florence (AL) Herald*, October 25, 1935, 1; "Plan Electrical Fair for County," *Limestone Democrat* November 5, 1936, 4.

53. "TVA Dinner Will Be Held Friday."

54. "Athens Is Praised by Lilienthal," *Alabama Courier* June 13, 1935, 7; "TVA Director Praises City," *Limestone Democrat* June 6, 1935, 1; "Lilienthal Pleased with Athens' Record," *Limestone Democrat* August 8, 1935, 1.

55. "Athens Honored in Radio Salute," *Alabama Courier* May 9, 1935, 8; "Rural Electrification: Again Athens Leads the Way!" *Limestone Democrat* May 30, 1935, 5; "Athens Congratulates Itself on the First Year of TVA Service," *Limestone Democrat* May 30, 1935, 8.

56. David E. Lilienthal, *The Journals of David E. Lilienthal, Volume I: The TVA Years* (New York: Harper & Row, 1964), 216.

57. Douglas B. Craig, *Fireside Politics: Radio and Political Culture in the Unit-*

ed States, 1920–1940 (Baltimore: Johns Hopkins University Press, 2005).

58. *Broadcasting & Cable Yearbook*, 1937, 49.

59. Joe Shirley, memo to Neil Savage, August 24, 1939, in TVA, Office of the General Manager, Information Office Correspondence Files, 1933–1946, RG 142 Box 14, National Archives–SERA.

60. Harcourt A. Morgan, "A National Program in the Tennessee Valley" (speech, Washington, DC, September 26, 1934), in TVA, Office of Engineering, Design, and Construction, Engineering Project Histories and Reports, RG 142 Box 23, National Archives–SERA.

61. David E. Lilienthal, radio address, January 1, 1936, in TVA, Office of the General Manager, Information Office Correspondence Files, 1933–1946, RG 142 Box 16, National Archives–SERA.

62. David E. Lilienthal, speech to Shelby County Young Democratic Club, Memphis, TN, October 20, 1934, in TVA, Office of the General Manager, Information Office Correspondence Files, 1933–1946, RG 142 Box 16, National Archives–SERA.

63. TVA Office of the General Manager, Information Office Correspondence Files, 1933–June 1944, RG 142 Box 7, page 12, National Archives–SERA.

64. "TVA News Reel at Ritz Friday Morning," *Limestone Democrat*, April 4, 1935, 1; "Free TVA Show at Palace Theatre next Sunday PM," *Guntersville Advertiser & Democrat*, June 26, 1935, 1.

65. "March of Time to Bring Impartial Story of TVA," *Limestone Democrat*, January 30, 1936, 2.

66. "March of Time to Bring Impartial Story of TVA."

67. "Model Kitchen to be Opened October 26," *Florence Herald*, October 10, 1935, 1.

68. Pare Lorentz, letter to Charles Krutch, Farm Security Administration, January 4, 1938, in TVA, Office of the General Manager, Information Office Correspondence Files, 1933–1946, RG 142 Box 13, National Archives–SERA.

69. TVA Office of the General Manager, Information Office Correspondence Files, 1933–June 1944, RG 142 Box 7, page 9, National Archives–SERA.

70. TVA, Office of Engineering, Design, and Construction, Engineering Project Histories and Reports, RG 142 Box 23, National Archives–SERA.

71. C. O. Gillingham, letter to W. L. Sturdevant, October 23, 1934, in TVA, Office of the General Manager, Information Office Correspondence Files, 1933–1946, RG 142 Box 6, National Archives–SERA.

72. C. O. Gillingham, letter to W. L. Sturdevant, October 23, 1934, in TVA, Office of the General Manager, Information Office Correspondence Files, 1933–1946, RG 142 Box 6, National Archives–SERA; TVA Office of the General Manager, Information Office, Correspondence Files, 1933–June 1944, RG 142 Box 1, National Archives–SERA.

73. TVA Office of the General Manager, Information Office Correspondence Files, 1933–June 1944, RG 142 Box 7, page 8, National Archives–SERA.

74. TVA, "Review of the Program and Policy of the Information Office," 7.

75. TVA Office of the General Manager, Information Office Correspondence Files, 1933–June 1944, RG 142 Box 7 Appendix 3, National Archives–SERA.

76. Mordecai Lee, *Congress vs. The Bureaucracy* (Norman: University of Oklahoma Press, 2011).

77. For a more complete description of New Deal agencies and public relations efforts, see Mordecai Lee, *Congress vs. The Bureaucracy.*

78. Charles Krutch, interview by Charles Crawford.

79. Charles Krutch, interview by Charles Crawford.

80. James Agee, "TVA I: Work in the Valley," *Fortune Magazine* (May 1935): 93.

81. Agee, "TVA I: Work in the Valley," 97.

82. Nichols, "Teaching Grandmother How to Spin."

Chapter 5

1. Barrett Shelton Sr., interview by Charles W. Crawford, June 18, 1970.

2. Barrett Shelton Sr., interview by Charles W. Crawford.

3. David E. Lilienthal, *The Journals of David E. Lilienthal,* 214–215.

4. Barrett Shelton Sr., Interview by Charles W. Crawford, June 18, 1970.

5. Lilienthal, *The Journals of David E. Lilienthal,* 214–215.

6. Lilienthal, *The Journals of David E. Lilienthal,* 214–215.

7. Paul Lazarsfeld, Bernard Berelson, and Hazel Gaudet, *The People's Choice,* 2nd ed. (New York: Columbia University Press, 1948).

8. Durisch, *Preliminary and Confidential Report–Wheeler,* 1935, 9.

9. "No New Dam News," *Guntersville Advertiser & Democrat,* September 3, 1935, 3.

10. "John Sparkman," *Limestone Democrat,* June 4, 1936, 8; "Pope Declares in State Chase on TVA Stand," *Chattanooga Daily Times,* January 5, 1934, 1.

11. "The Nagging Continues," *Limestone Democrat,* September 20, 1934, 4.

12. Maxwell E. McCombs and Donald L. Shaw, "The Agenda-Setting Function of Mass Media," *Public Opinion Quarterly* 36, no. 2 (1972): 176–187.

13. TVA, Office of the General Manager, Information Office, Correspondence Files, 1933–June 1944, RG 142 Box 1, National Archives–SERA.

14. TVA, Office of the General Manager, Information Office, Correspondence Files, 1933–June 1944, RG 142 Box 1, National Archives–SERA.

15. Thomas K. McCraw, *TVA and the Power Fight,* 148.

16. R. L. Duffus, letter to W. L. Sturdevant, January 13, 1938, in TVA, Office of the General Manager, Information Office Correspondence Files, 1933–1946, RG 142 Box 9, National Archives–SERA.

17. R. L. Duffus, letter to W. L. Sturdevant, May 13, 1936, in TVA, Office of the General Manager, Information Office Correspondence Files, 1933–1946, RG 142 Box 2, National Archives–SERA.

18. W. L. Sturdevant, letter to R. L. Duffus, April 4, 1936, in TVA, Office of the General Manager, Information Office Correspondence Files, 1933–1946, RG 142 Box 2, National Archives–SERA.

19. W. L. Sturdevant, letter to R. L. Duffus, April 29, 1936, in TVA, Office of the General Manager, Information Office Correspondence Files, 1933–1946, RG 142 Box 2, National Archives–SERA.

20. R. L. Duffus, "A Dream Takes Form in TVA's Domain," *New York Times*, April 19, 1936, SM8.

21. R. L. Duffus later went on to publish a book about the TVA with photographer Charles Krutch titled *A Valley and Its People*.

22. George Slover, letter to W. L. Sturdevant, March 10, 1037, in TVA, Office of the General Manager, Information Office Correspondence Files, 1933–1946, RG 142 Box 9, National Archives–SERA.

23. W. L. Sturdevant, letter to R. L. Duffus, January 7, 1937, in TVA, Office of the General Manager, Information Office Correspondence Files, 1933–1946, RG 142 Box 2, National Archives–SERA.

24. John Gunther, *The Story of TVA* (New York: Harper, 1947), 1. The author did not indicate which three newspaper editors were unfavorable toward the TVA.

25. Fannon Beauchamp, interview by Charles W. Crawford.

26. R. L. Duffus, "Big Dams to Change the Face of America," *New York Times* September 8, 1935, E11.

27. "Governor Says South Blessed by TVA Work," *Chattanooga Daily Times*, April 10, 1934, 1.

28. "Word Picture of Modern Utopia Painted," *Decatur Daily*, March 7, 1934, 7.

29. "Tennessee is Truly River of Romance," *Guntersville Advertiser & Democrat*, February 8, 1933, 1; "Writer Describes Fertile Tennessee Valley," *Guntersville Advertiser & Democrat*, March 15, 1933, 4; "Tennessee, Lazy Old River," *Guntersville Advertiser & Democrat*, August 2, 1933, 4.

30. "Will Marshall County Develop?" *Albertville Herald*, July 30, 1936, 2.

31. "Tennessee Valley Authority Chairman Makes Statement," *Guntersville Advertiser & Democrat*, August 2, 1933, 1.

32. "Tennessee Valley Authority Bringing New Social Order to the Mountains," *Decatur Daily*, March 8, 1934, 5.

33. "Tennessee Valley Authority Bringing New Social Order to the Mountains."

34. "Three Executives in Charge of Uncle Sam's Development."

35. "Three Executives in Charge of Uncle Sam's Development."

36. "Employment on Norris Dam Given," *Atlanta Daily World*, April 24, 1934, 1; "TVA To Work Both Races Equally," *Atlanta Daily World*, June 20, 1934, 1.

37. "Lilienthal Set Power Bargain that Hits City," *Chattanooga Daily Times*, April 6, 1934, 1; "New Authority to be Directed from this City," *Chattanooga Daily Times*, January 1, 1934, 1.

38. R. L. Duffus, "A Dream Takes Form in TVA's Domain."

39. W. O. Foster, "Tennessee Valley Wakes from Dream: Finds the New Deal

Plan is not Federal Dole in Disguise," *New York Times*, December 3, 1933.

40. "TVA Officials Visit Athens," *Limestone Democrat*, November 1, 1934, 1.

41. "Alabama's Rights," *Limestone Democrat*, November 28, 1935, 4.

42. "Gets 40 Percent Reduction under New TVA Contract" *Alabama Courier*, June 14, 1934, 2; "Lauderdale Farmers Use TVA Power," *Limestone Democrat*, November 8, 1934, 5; "Pulaski Is First to Get TVA Power," *Limestone Democrat*, January 10, 1935, 5; "City of Tupelo, MS Electricity Department Income Account," *Limestone Democrat*, March 14, 1935, 4; "Pulaski Saves $14,000 by Using TVA Power," *Limestone Democrat*, January 2, 1936, 8.

43. "Nashville Sees Opportunity," *Limestone Democrat*, September 27, 1934, 4.

44. "Drink Dr. Pepper," *Alabama Courier*, August 30, 1934, 4.

45. "The Tennessee Valley Authority and the Public Service Commission," *Albertville Herald*, July 19, 1934, 2.

46. "Huge TVA Project: Millions Available as Soon as Senate and President OK Action of House," *Albertville Herald*, June 27, 1935, 1.

47. "FDR Is Lauded for TVA Deal," *Albertville Herald*, March 17, 1936, 4.

48. "Fight Is On," *Alabama Courier*, September 24, 1936, 4.

49. "Fight Is On."

50. "FDR Is Lauded for TVA Deal."

51. "What Would Washington Do under Present Conditions?" *Albertville Herald*, February 20, 1936, 2.

52. "The Value of TVA Power," *Limestone Democrat*, July 4, 1935, 1.

53. "The Value of TVA Power."

54. "TVA Power Contract Is Signed by Decatur." *Decatur Daily*, March 16, 1934, 1; "City Council Declares Election Carried for TVA," *Albertville Herald*, September 24, 1937, 1; "Ft. Payne Votes for TVA Power System," *Albertville Herald*, November 26, 1937, 1.

55. "Local Power Plans," *Florence Herald*, May 22, 1936, 1.

56. "New Deal Foes Will Get Skids, Farley Claims," *Chattanooga Daily Times*, January 14, 1934, 1.

57. "TVA Officials Visit Athens."; "Lilienthal Visits Athens," *Alabama Courier*, December 30, 1934, 1.

58. "FDR Visits Alabama," *Albertville Herald*, November 22, 1934, 1.

59. "Thousands Greet FDR in Athens," *Limestone Democrat*, November 22, 1934, 1.

60. "Representative of Tennessee Valley Authority Visits Guntersville Seeking Labor," *Guntersville Advertiser & Democrat*, September 30, 1933, 1; "Examinations for TVA Jobs," *Guntersville Advertiser & Democrat*, December 6, 1933, 4; "Examination for TVA Labor to be Held This Month," *Guntersville Advertiser & Democrat*, January 1, 1936, 1; "Authority Gives 2nd Examination," *Limestone Democrat*, January 2, 1936, 8; "TVA to Give Tests for Jobs on Dams," *Florence Herald*, January 3, 1936, 1; "Ample TVA Blanks Will Be Given Out," *Limestone Democrat*, January 9, 1936, 1.

61. "TVA Reservoirs Division Opens Office in Decatur," *Decatur Daily*, February 22, 1934, 1.

62. "Big Developments Promised for the Tennessee Valley," *Guntersville Advertiser & Democrat*, January 25, 1933, 1.

63. "Limiting Production Foolhardy," *Chattanooga Daily Times*, April 25, 1934, 1.

64. "Morgan to Develop Tennessee Valley: Chairman of Federal Authority Will Seek to Preserve Permanent Agriculture." *New York Times*, June 8, 1933.

65. R. L. Duffus, "Jobs for Four Million in Place of Relief," *New York Times*, November 19, 1933, 1.

66. R. L. Duffus, "A Dream Takes Form in TVA's Domain."

67. "Attention of Home Builders: TVA Dam and Other Prospective Developments Turn Eyes on Albertville," *Albertville Herald*, August 8, 1935, 1; "Huntsville Expects Trade Increase from Dam," *Guntersville Advertiser & Democrat*, December 18, 1935, 2.

68. "TVA Urged to Bring Industries to District," *Guntersville Advertiser & Democrat*, November 6, 1935, 1.

69. "Tupelo Power Bill Is Lower," *Alabama Courier*, February 9, 1934, 1.

70. "Alabama Ranks Low in Its Electrification," *Limestone Democrat*, August 20, 1936, 4.

71. "Writer Describes Fertile Tennessee Valley Roosevelt Plans to Develop," *Guntersville Advertiser & Democrat*, March 15, 1933, 4.

72. "Power Tells the Story," *Guntersville Advertiser & Democrat*, February 15, 1933, 1.

73. "56 Families Use over 400 KWH per Month," *Limestone Democrat*, August 15, 1935, 8.

74. "Rural Electrification: Again Athens Leads the Way!" *Limestone Democrat*, May 30, 1935.

75. "New Era of Power Revolutionizes Life in the Tennessee Valley," *New York Times*, November 29, 1936.

76. Russell B. Porter, "TVA Power Lifts Farm Standards," *New York Times*, April 25 1938, 21.

77. "New Era of Power Revolutionizes Life in the Tennessee Valley," *New York Times*, November 29, 1936.

78. "Looks Like Everybody's Buying a General Electric Refrigerator!" *Limestone Democrat*, September 3, 1936, 1.

79. "9,785 Refrigerators Sold in Alabama in 1933," *Huntsville Times*, February 4, 1934, 1.

80. "Sale of Electric Items Increases," *Florence Herald*, September 25, 1936, 1.

81. "Sale of Electric Items Increases," *Florence Herald*; "Florence Now Buying Electrical Appliances," *Limestone Democrat*, October 8, 1936, 2.

82. "Athens Heads List of Range Users," *Limestone Democrat*, April 25, 1935, 1; "Athens Leads in Electricity Use," *Limestone Democrat*, May 23, 1935, 1.

83. "Appliance Sale Continues Climb," *Limestone Democrat*, May 30, 1935, 1.

84. "U. G. White Hardware," *Limestone Democrat*, April 11, 1935, 4.

85. "Sterchis." *Huntsville Times*, January 2, 1934, 3.

86. "Now Is the Time to Have Your Electric Appliance Installed," *Limestone Democrat*, June 13, 1935, 1.

87. "Government to Assist Electric Appliance Purchasers," *Guntersville Advertiser & Democrat*, December 27, 1933, 1.

88. "TVA Is Our Papa," *Alabama Courier*, February 20, 1936, 8.

89. "Electrolux Tempted?" *Limestone Democrat*, September 24, 1936, 1.

90. "Be Ready—Prepare for the TVA," *Alabama Courier*, April 19, 1934, 3.

91. "TVA: Electricity for All," *Alabama Courier*, November 8, 1934, 7.

92. "TVA: Electricity for All," *Alabama Courier*, November 15, 1934, 8.

93. Family case record, "Lucas, Henry," TVA Population Removal Records, Family Relocation Files 1934–1954, RG 142 Box 67, National Archives–SERA; Family case record, "Cagle, W. P.," TVA Population Removal Records, Family Relocation Files 1934–1954, RG 142 Box 78, National Archives–SERA.

94. "Athens Earns $6,500 TVA Power First Year," *Limestone Democrat*, September 5, 1935, 6.

95. "City of Tupelo, MS Electricity Department Income Account," *Limestone Democrat*, March 14, 1935, 4.

96. "Pulaski Is First to Get TVA Power," *Limestone Democrat*, January 10, 1935, 5.

97. "Homes at Norris Are to be Entirely Electrified," *Decatur Daily*, January 23, 1934, 1.

98. "Homes at Norris Are to be Entirely Electrified," *Decatur Daily*.

99. "Re-locating Meeting to be Held in Guntersville," *Guntersville Advertiser & Democrat*, March 11, 1936, 1.

100. "Families in Guntersville Dam Area to be Assisted in Getting Re-located," *Guntersville Advertiser & Democrat*, March 12, 1936, 1.

101. "Plans for Re-locating Farm Families to be Discussed at Farm Meeting," *Albertville Herald*, May 7, 1936, 1; "Plans for Relocating Farm Families to be Discussed at Meetings," *Guntersville Advertiser & Democrat*, May 6, 1936, 1; "Assistance Available for Farm Families in Relocation," *Albertville Herald*, April 30, 1936, 1.

102. "Tennessee Valley Landowners Mutual Aid Agency," *Limestone Democrat*, December 13, 1934, 8; "Tennessee Valley Landowners Mutual Aid Agency," *Alabama Courier*, December 13, 1934, 8; "Tennessee Valley Landowners Mutual Aid Agency," *Alabama Courier*, December 20, 1934, 10; "Tennessee Valley Landowners Mutual Aid Agency," *Guntersville Advertiser & Democrat*, January 30, 1935, 2; "Tennessee Valley Landowners Mutual Aid Agency," *Florence Herald*, January 25, 1936, 1.

103. Downs, *Transforming the South*.

104. "Tennessee Valley Landowners Mutual Aid Agency," *Florence Herald*.

105. "Shoals Dam Work Delayed by Caves; Discovery of Network under Cove Creek Area Worries Government Officials," *New York Times*, July 11, 1933, 25.

106. "TVA Plans Change in Life of District: Morgan Says Local Industries Will Be Used to Draw Some People from Farms," *New York Times*, May 11, 1934.

107. David E. Lilienthal, letter to W. L. Sturdevant, January 22, 1935, in TVA Office of the General Manager, Records of the Board of Directors, David E. Lilienthal General Correspondence, 1933–1946, RG 142 Box 4, in National Archives–SERA.

108. David E. Lilienthal, February 20, 1937, in TVA Office of the General Manager, Records of the Board of Directors, David E. Lilienthal General Correspondence, 1933–1946, RG 142 Box 1, in National Archives–SERA.

109. Letter from C. M. Stanley to David E. Lilienthal, January 29, 1936, in Office of Information files, RG 142, Box 4, National Archives–SERA.

110. Letter from W. L. Sturdevant to "editor" at Mobile Register, n.d., in Office of Information files, RG 142, Box 4, National Archives–SERA.

111. W. G. Foster, "Tennessee Valley Annoyed by Delay," *New York Times*, September 17, 1933.

112. Charles Puckett, "New Horizons in the TN Valley: What the Conservative People of the Region Think of the Government's Great Experiment in Social Planning" *New York Times*, November 18, 1934.

113. Agee, "TVA I: Work in the Valley," 146.

114. Agee, "TVA I: Work in the Valley," 146.

115. "Protests Site of TVA," *Huntsville Times*, January 2, 1934, 1.

116. "Norris Dam and New City," *Albertville Herald*, April 12, 1934, 3.

117. "The Ontario Fiasco," *Florence Herald*, May 3, 1935, 1.

118. "After Payroll Goes—What?" *Florence Herald*, August 2, 1935, 1.

119. TVA, *Report to the Congress on the Unified Development of the Tennessee River System*, 1936, 38.

Chapter 6

1. R. L. Duffus, *The Valley and Its People: A Portrait of TVA*, 60.

2. Charles Kenneth Roberts, "New Deal Community-building in the South: The Subsistence Homesteads around Birmingham, Alabama." *The Alabama Review* 66, no. 2, 83–121.

3. Family case record, "Bennett John," TVA Population Removal Records, Family Relocation Files 1934–1954, RG 142 Box 78, National Archives–SERA.

4. Family case record, "Hastings, Marvin," TVA Population Removal Records, Family Relocation Files 1934–1954, RG 142 Box 80, National Archives–SERA.

5. Luther Tidwell, interview by Laura Beth Daws.

6. Family case record, "Baker, Carver (Box 51)"; "Bowens, Sim (Box 51)";

"Hopper, Mrs. William (Box 54)," TVA Population Removal Records, Family Relocation Files 1934–1954, RG 142, National Archives–SERA.

7. Hazel Moore Thompson, interview by Laura Beth Daws.

8. Hazel Moore Thompson, interview by Laura Beth Daws.

9. Family case record, "Ross, Gus," TVA Population Removal Records, Family Relocation Files 1934–1954, RG 142 Box 81, National Archives–SERA.

10. Family case record, "Walker, Jeff," TVA Population Removal Records, Family Relocation Files 1934–1954, RG 142 Box 82, National Archives–SERA.

11. Hazel Moore Thompson, interview by Laura Beth Daws.

12. Family case record, "Brown, McKinley," TVA Population Removal Records, Family Relocation Files 1934–1954, RG 142 Box 78, National Archives–SERA.

13. Family case record, "Gilliam, George," TVA Population Removal Records, Family Relocation Files 1934–1954, RG 142 Box 79, National Archives–SERA.

14. Family case record, "Hilburn, Charlie," TVA Population Removal Records, Family Relocation Files 1934–1954, RG 142 Box 80, National Archives–SERA.

15. Family case record, "Grant, C. M.," TVA Population Removal Records, Family Relocation Files 1934–1954, RG 142 Box 80, National Archives–SERA.

16. Family case record, "Cloud, Charlie."

17. Family case record, "Toney, Pat," TVA Population Removal Records, Family Relocation Files 1934–1954, RG 142 Box 58, National Archives–SERA.

18. Family case record, "Toney, Pat."

19. Family case record, "Culbert, Joe," TVA Population Removal Records, Family Relocation Files 1934–1954, RG 142 Box 79, National Archives–SERA.

20. Family case record, "Andrews, John," TVA Population Removal Records, Family Relocation Files 1934–1954, RG 142 Box 78, National Archives–SERA.

21. Family case record, "Shelton, James," TVA Population Removal Records, Family Relocation Files 1934–1954, RG 142 Box 82, National Archives–SERA.

22. Family case record, "Mason, Jim," TVA Population Removal Records, Family Relocation Files 1934–1954, RG 142 Box 67, National Archives–SERA.

23. Family case record, "Legg, Earl," TVA Population Removal Records, Family Relocation Files 1934–1954, RG 142 Box 66, National Archives–SERA.

24. Family case record, "Rice, Will," TVA Population Removal Records, Family Relocation Files 1934–1954, RG 142 Box 81, National Archives–SERA.

25. Family case record, "Youngblood, Edmund" TVA Population Removal Records, Family Relocation Files 1934–1954, RG 142 Box 59, National Archives–SERA.

26. Family case record, "Burnham, Mary," TVA Population Removal Records, Family Relocation Files 1934–1954, RG 142 Box 63, National Archives–SERA.

27. Family case record, "Auston, William," TVA Population Removal Re-

cords, Family Relocation Files 1934–1954, RG 142 Box 81, National Archives–SERA.

28. Family case record, "Williams, Mary Jane," TVA Population Removal Records, Family Relocation Files 1934–1954, RG 142 Box 82, National Archives–SERA.

29. "The Tennessee Valley Authority and the Public Service Commission," *Albertville Herald*, July 19, 1934, 2.

30. "Redbook on TVA," *Alabama Courier*, July 11, 1935, 2.

31. "TVA Fight Is Just Started," *Alabama Courier*, July 11, 1935, 1.

32. "Redbook Magazine Article Arouses Anger of Athens," *Limestone Democrat*, July 11, 1935, 1.

33. "Redbook on TVA," *Alabama Courier*, 2.

34. "Citizens Meet to Endorse TVA," *Limestone Democrat*, July 25, 1935, 1.

35. "Citizens Meet to Endorse TVA," *Limestone Democrat*.

36. "Article on TVA is Rapped by Athens," *Limestone Democrat*, July 18, 1935, 5; "Redbook Magazine Article Arouses Anger of Athens," *Limestone Democrat*, July 11, 1935, 1.

37. "Redbook Magazine Article Arouses Anger of Athens," *Limestone Democrat*.

38. Family case record, "Woods, Rister (Box 58)"; "Hammond, J. D. (Box 53)"; "Adcock, T. J. (Box 62)"; and "Hargiss, Albert," TVA Population Removal Records, Family Relocation Files 1934–1954, RG 142, National Archives–SERA.

39. Family case record, "Abernathy, Dick," TVA Population Removal Records, Family Relocation Files 1934–1954, RG 142 Box 50, National Archives–SERA.

40. Family case record, "Carter, Millard," TVA Population Removal Records, Family Relocation Files 1934–1954, RG 142 Box 79, National Archives–SERA.

41. Family case record, "Higdon, Andrew," TVA Population Removal Records, Family Relocation Files 1934–1954, RG 142 Box 65, National Archives–SERA.

42. Family case record, "Flanagan, Luke," TVA Population Removal Records, Family Relocation Files 1934–1954, RG 142 Box 64, National Archives–SERA.

43. Family case record, "Bevins, Albert," TVA Population Removal Records, Family Relocation Files 1934–1954, RG 142 Box 51, National Archives–SERA.

44. Family case record, "Moore, Fred," TVA Population Removal Records, Family Relocation Files 1934–1954, RG 142 Box 67, National Archives–SERA.

45. Family case record, "Andrews, Elijah."

46. Family case record, "Gilliland, Ada," TVA Population Removal Records, Family Relocation Files 1934–1954, RG 142 Box 79, National Archives–SERA.

47. Eugene Simonson, interview by Laura Beth Daws, July 19, 2012.

48. TVA intended for these work camps to be family-friendly areas away from a number of diversions, such as alcohol, prostitutes, and gambling. How-

ever, Slab Town in the Pickwick Dam reservoir area arose to offer employees these and other distractions. For a more thorough discussion of damtowns and their competition, see Steve Killingsworth, "Regional Planning on a Local Scale: The Brief Life of Pickwick Village," *West Tennessee Historical Society Papers* 51 (1997): 48–63.

49. Hardin quotes: Interview by Daws; Medical issues: Family case record, "Fuqua, Susie," TVA Population Removal Records, Family Relocation Files 1934–1954, RG 142 Box 53, National Archives–SERA.

50. Family case record, "Dunn, John P.," TVA Population Removal Records, Family Relocation Files 1934–1954, RG 142 Box 79, National Archives–SERA.

51. Family case record, "Blackwood, Joe," TVA Population Removal Records, Family Relocation Files 1934–1954, RG 142 Box 51, National Archives–SERA.

52. Family case record, "Hooper, Bess Keeton," TVA Population Removal Records, Family Relocation Files 1934–1954, RG 142 Box 80, National Archives–SERA; Family case record, "Keasling, J. A.," TVA Population Removal Records, Family Relocation Files 1934–1954, RG 142 Box 80, National Archives–SERA.

53. Family case record, "Allen, Walter," TVA Population Removal Records, Family Relocation Files 1934–1954, RG 142 Box 78, National Archives–SERA.

54. Family case record, "Green, Thomas."

55. Family case record, "Langston, John," TVA Population Removal Records, Family Relocation Files 1934–1954, RG 142 Box 55, National Archives–SERA.

56. Family case record, "Sheffield, Vernon," TVA Population Removal Records, Family Relocation Files 1934–1954, RG 142 Box 82, National Archives–SERA.

57. Family case record, "Rice, Dave, Jr."

58. Family case record, "Bowling, J. P.," TVA Population Removal Records, Family Relocation Files 1934–1954, RG 142 Box 78, National Archives–SERA.

59. Family case record, "Bowling, J. P.," TVA Population Removal Records, Family Relocation Files 1934–1954, RG 142 Box 78, National Archives–SERA.

60. Family case record, "Bowling, J. P."

61. Family case record, "Bowling, J. P."

62. Bill Hardin, interview by Laura Beth Daws.

63. Family case record, "Fitchead, Elias," TVA Population Removal Records, Family Relocation Files 1934–1954, RG 142 Box 53, National Archives–SERA.

64. Family case record, "Baty, John," TVA Population Removal Records, Family Relocation Files 1934–1954, RG 142 Box 51, National Archives–SERA.

65. Family case record, "Scoggin, Hugh," TVA Population Removal Records, Family Relocation Files 1934–1954, RG 142 Box 68, National Archives–SERA.

66. Family case record, "Blackwell, Tom," TVA Population Removal Records, Family Relocation Files 1934–1954, RG 142 Box 62, National Archives–SERA.

67. Family case record, "Arnel, Frank," TVA Population Removal Records, Family Relocation Files 1934–1954, RG 142 Box 78, National Archives–SERA.

68. Family case record, "Hapton, Waymon," TVA Population Removal Re-

cords, Family Relocation Files 1934–1954, RG 142 Box 80, National Archives–SERA.

69. Paul Conner, interview by Laura Beth Daws and Susan Brinson.

70. Family case record, "Hapton, Waymon."

71. TVA, *Preliminary and Confidential Report–Wheeler*, 1935.

72. TVA, *Population Readjustment, Guntersville Area*, 15.

73. William E. Leuchtenburg, *Franklin D. Roosevelt and the New Deal, 1932–1940* (New York: Harper & Row, 1963), 185.

74. Aaron D. Purcell, *White Collar Radicals*, 10.

75. Family case record, "Hapton, Waymon."

76. Family case record, "Hayes, Robert," TVA Population Removal Records, Family Relocation Files 1934–1954, RG 142 Box 80, National Archives–SERA.

77. Family case record, "Points, Clarence," TVA Population Removal Records, Family Relocation Files 1934–1954, RG 142 Box 56, National Archives–SERA.

78. Family case record, "Points, Clarence."

79. Family case record, "Horton, Ernest," TVA Population Removal Records, Family Relocation Files 1934–1954, RG 142 Box 80, National Archives–SERA.

80. "Family Case Record, "Anderson, Lucy," TVA Population Removal Records, Family Relocation Files 1934–1954, RG 142 Box 58, National Archives–SERA. "Unceiled" means that the structure had no ceiling.

81. Family case record, "Horton, Ernest."

82. Family case record, "Horton, Henry," TVA Population Removal Records, Family Relocation Files 1934–1954, RG 142 Box 80, National Archives–SERA.

83. Family case record, "Merrill, Bud," TVA Population Removal Records, Family Relocation Files 1934–1954, RG 142 Box 81, National Archives–SERA.

84. Family case record, "Scott, L. W.," TVA Population Removal Records, Family Relocation Files 1934–1954, RG 142 Box 81, National Archives–SERA.

85. Family case record, "Chandler, Henry," TVA Population Removal Records, Family Relocation Files 1934–1954, RG 142 Box 79, National Archives–SERA.

86. Family case record, "Ruffin, Ernest," TVA Population Removal Records, Family Relocation Files 1934–1954, RG 142 Box 68, National Archives–SERA.

87. Family case record, "Griffin, I. C.," TVA Population Removal Records, Family Relocation Files 1934–1954, RG 142 Box 80, National Archives–SERA.

88. Grant, *TVA and Black Americans*.

89. C. H. Pritchett and C. M. Stephenson, *Preliminary and Confidential Report–Guntersville*, 1935, 1.

90. Pritchett and Stephenson, *Preliminary and Confidential Report–Guntersville*, 3.

91. Durisch, *Preliminary and Confidential Report Wheeler*, 17.

92. TVA, "The Wheeler Project Technical Report 2," 1939, 256.

93. *Family Relocation Analysis: Norris, Guntersville, Wheeler*, in TVA Res-

ervoir Property Management, Population Removal Records, Administrative Files, RG 142 Box 6.

94. Family case record, "Hutcheson, Joe," TVA Population Removal Records, Family Relocation Files 1934–1954, RG 142 Box 80, National Archives–SERA.

95. Family case record, "Huggins, Harrison," TVA Population Removal Records, Family Relocation Files 1934–1954, RG 142 Box 80, National Archives–SERA.

96. Family case record, "Hester, B. L.," TVA Population Removal Records, Family Relocation Files 1934–1954, RG 142 Box 80, National Archives–SERA.

97. Family case record, "Edmonds, Cleve," TVA Population Removal Records, Family Relocation Files 1934–1954, RG 142 Box 79, National Archives–SERA.

98. Paul Conner, interview by Laura Beth Daws and Susan Brinson.

99. Letter from W. G. Carnahan, September 1, 1938, in Family case record, "Fillmore, Willie," TVA Population Removal Records, Family Relocation Files 1934–1954, RG 142 Box 79, National Archives–SERA.

100. Family case record, "Pegues, Fred," TVA Population Removal Records, Family Relocation Files 1934–1954, RG 142 Box 81, National Archives–SERA.

101. Family case record, "Bell, Jim," TVA Population Removal Records, Family Relocation Files 1934–1954, RG 142 Box 78, National Archives–SERA.

102. Family case record, "Elledge, Oscar," TVA Population Removal Records, Family Relocation Files 1934–1954, RG 142 Box 79, National Archives–SERA.

103. Family case record, "McCrary, Bob," TVA Population Removal Records, Family Relocation Files 1934–1954, RG 142 Box 81, National Archives–SERA.

104. TVA, *Population Readjustment, Guntersville Area*, 11.

105. TVA, *Population Readjustment in the Tennessee Valley*, 1945.

106. TVA, *Population Readjustment in the Tennessee Valley*.

107. Family case record, "Baker, Herman," TVA Population Removal Records, Family Relocation Files 1934–1954, RG 142 Box 78, National Archives–SERA.

108. Family case record, "Nelson, Nancie," TVA Population Removal Records, Family Relocation Files 1934–1954, RG 142 Box 56, National Archives–SERA.

109. Family case record, "Elrod, Alex," TVA Population Removal Records, Family Relocation Files 1934–1954, RG 142 Box 54, National Archives–SERA.

110. Family case record, "Moore, John," TVA Population Removal Records, Family Relocation Files 1934–1954, RG 142 Box 57, National Archives–SERA.

Conclusion

1. Luther Tidwell, interview by Laura Beth Daws.

2. Family case record, "Ausbon, Lee," TVA Population Removal Records, Family Relocation Files 1934–1954, RG 142 Box 50, National Archives–SERA.

3. Bill Hardin, interview by Laura Beth Daws.

4. Maxine Williamson Black, interview by Laura Beth Daws, June 7, 2012.

5. Beryl Tidwell, interview by Laura Beth Daws.

6. Gunther, *Story of TVA*, 1.

7. Owen, *The Tennessee Valley Authority* (New York: Praeger, 1973), 217.

8. TVA, *The Wheeler Project Technical Report 2*, 283; TVA, *The Guntersville Technical Report 4*, 301.

9. TVA, *The Wheeler Project Technical Report 2*, 1939, 283; TVA, *The Guntersville Technical Report 4*, 301.

10. Chandler, *Myth of TVA*, 57.

11. TVA, *The Wheeler Project Technical Report 2*, 1939, 256.

12. TVA, *The Wheeler Project Technical Report 2*.

13. TVA, *The Wheeler Project Technical Report 2*.

14. Family case record, "Boldin, Sam," TVA Population Removal Records, Family Relocation Files 1934–1954, RG 142 Box 62, National Archives–SERA.

15. Family case record, "Hardin, Tom," TVA Population Removal Records, Family Relocation Files 1934–1954, RG 142 Box 80, National Archives–SERA.

16. TVA, *The Wheeler Project Technical Report 2*, 11.

17. US Census Bureau, 1940 census.

18. Family case record, "Orr, Pleas," TVA Population Removal Records, Family Relocation Files 1934–1954, RG 142 Box 67, National Archives–SERA.

19. Family case record, "Pack, Herbert," TVA Population Removal Records, Family Relocation Files 1934–1954, RG 142 Box 68, National Archives–SERA.

20. TVA, *The Wheeler Project Technical Report 2*, 1939, 256.

21. Family case record, "Panel, Jess," TVA Population Removal Records, Family Relocation Files 1934–1954, RG 142 Box 81, National Archives–SERA.

22. US Census Bureau, 1940 census.

23. TVA, *Population Readjustment in the Tennessee Valley*, 1945.

24. Gunther, *Story of TVA*, 4.

25. Mary Utopia Rothrock, interview by Charles Crawford.

26. Gunther, *Story of TVA*, 4.

27. Earle Sumner Draper, interview by Charles W. Crawford.

28. Paul Conner, interview by Laura Beth Daws and Susan Brinson.

29. US Census Bureau, 1940 census.

30. Paul Conner, interview by Laura Beth Daws and Susan Brinson.

31. Eugene Simonson, interview by Laura Beth Daws.

32. US Census Bureau, 1940 census.

33. Bill Hardin, interview by Laura Beth Daws.

34. Wilson Whitman, *God's Valley*.

35. T. M. N. Lewis, "Indian Mounds Now Covered by Lake Tell the Story," article clipped from an unknown newspaper, in TVA Office of Engineering, Design, and Construction, Engineering Project Histories and Reports, RG 142 Box 89, National Archives–SERA.

36. Gunther, *Story of TVA*, 1.

37. Thomas Martin, *The Story of Electricity in AL since the Turn of the Century, 1900–1952* (Birmingham: Alabama Power Company, 1952), 112.

38. Owen, *The Tennessee Valley Authority*, 215.

39. Eugene Simonson, interview by Laura Beth Daws.

40. David E. Lilienthal, *Democracy on the March*.

41. Paul Evans, interview by Charles W. Crawford, August 7, 1979.

42. TVA, *Report to the Congress on the Unified Development of the Tennessee River System*, 1936, 38.

43. Bill Hardin, interview by Laura Beth Daws.

44. Beryl Tidwell, interview by Laura Beth Daws.

45. Hazel Moore Thompson, interview by Laura Beth Daws.

46. Dixie Conner, interview by Laura Beth Daws and Susan Brinson, August 6, 2012.

47. Maxine Williamson Black, interview by Laura Beth Daws, June 7, 2012.

Bibliography

Adamson, W. M. *Income in Counties of Alabama, 1929–1935*. Tuscaloosa: Bureau of Business Research, School of Commerce and Business Administration, University of Alabama, 1939.

Agee, James. "TVA I: Work in the Valley." *Fortune Magazine*, May 1935.

Agee, James, and Walker Evans. *Let Us Now Praise Famous Men*. Boston: Houghton Mifflin, 1941.

Alldredge, J. Haden, Mildred Burnham Spottswood, Vera V. Anderson, John H. Goff, and Robert M. LaForge. *A History of the Navigation on the Tennessee River System*. Washington, DC: Transportation Economics Division, Tennessee Valley Authority, 1937.

Annual Report of the Tennessee Valley Authority for the Fiscal Year ended June 30 1935. New York: Arno Press, 1969.

Atkins, Leah Rawls. *Developed for the Service of Alabama: The Centennial History of the Alabama Power Company, 1906–2006*. Birmingham: Alabama Power Company, 2006.

Badger, Anthony. *New Deal/New South*. Fayetteville: University of Arkansas Press, 2007.

Barker, Bessie. "A Study of the Living Conditions of One Hundred Families in Limestone County, Alabama on Various Economic Levels." Master's thesis, Alabama Polytechnic Institute, 1934.

Beauchamp, Fannon. Interview by Charles W. Crawford, June 16, 1970.

Biles, Roger. *The South and the New Deal*. Lexington: University Press of Kentucky, 1994.

Black, Maxine Williamson. Interview by Laura Beth Daws, June 17, 2012.

Broadcasting & Cable Yearbook. Washington, DC: Broadcasting Publications, Inc., 1937.

Brown, James Sea, Jr., ed. *Up before Daylight: Life Histories from the Alabama Writers' Project, 1938–1939*. Tuscaloosa: University of Alabama Press, 1982.

Chandler, William U. *The Myth of TVA: Conservation and Development in the Tennessee Valley, 1933–1983*. Cambridge, MA: Ballinger Publishing Company, 1984.

Clapp, Gordon R. *The TVA: An Approach to the Development of a Region*. Chicago: University of Chicago Press, 1955.

Conner, Paul and Dixie. Interview by Laura Beth Daws and Susan Brinson, August 6, 2012.

Conner, T. L. Interview by Laura Beth Daws, July 19, 2012.

Conrad, David E. *The Forgotten Farmers: The Story of Sharecroppers in the New Deal*. Urbana: University of Illinois Press, 1965.

Craig, Douglas B. *Fireside Politics: Radio and Political Culture in the United States, 1920–1940*. Baltimore: Johns Hopkins University Press. 2005.

Curry, Bobbie. Interview by Laura Beth Daws, July 19, 2012.

Daniel, Pete. "The New Deal, Southern Agriculture, and Economic Change." In *The New Deal and the South*, ed. James C. Cobb and Michael V. Namorato, 37–61. Jackson: University Press of Mississippi, 1984.

Dawson, J. Dudley, and Arthur E. Morgan. Interview by Charles W. Crawford, June 20, 1969.

Downs, Matthew. *Transforming the South: Federal Development in the Tennessee Valley, 1915–1960*. Baton Rouge: Louisiana State University Press, 2014.

Draper, Earle Sumner. Interview by Charles W. Crawford, December 30, 1969.

Duffus, R. L. *The Valley and Its People: A Portrait of TVA*. New York: A. A. Knopf, 1946.

Durisch, Lawrence. *Preliminary and Confidential Report: Families of the Wheeler Reservoir Area*. TVA, Research Section, Social and Economic Division. Knoxville, TN: TVA, September 12, 1935.

Evans, Paul. Interview by Charles W. Crawford, August 7, 1979.

Falck, Edward. Interview by Charles W. Crawford, September 25, 1970.

Field, Gregory B. "'Electricity for All': The Electric Home and Farm Authority and the Politics of Mass Consumption, 1932–1935." *The Business History Review* 64, no. 1 (1990): 32–60.

Flynt, Wayne. *Alabama in the 20th Century*. Tuscaloosa: University of Alabama Press, 2004.

———. *Poor but Proud: Alabama's Poor Whites*. Tuscaloosa: University of Alabama Press, 1989.

Freidel, Frank. "The South and the New Deal." In *The New Deal and the South*, edited by James C. Cobb and Michael V. Namorato, 17–36. Jackson: University Press of Mississippi, 1984.

Grant, Nancy. *TVA and Black Americans: Planning for the Status Quo*. Philadelphia: Temple University Press, 1990.

Grantham, Dewey W. "Regional Claims and National Purposes: The South and the New Deal." *Atlanta History: A Journal of Georgia and the South* 38, no. 3 (1994): 5–17.

Gray, Daniel Savage, J. Barton Starr, and Linda Crockett Gray. *Alabama: A Place, A People, A Point of View*. Dubuque, IA: Kendall/Hunt Pub. Co., 1977.

Grubbs, Donald H. *Cry from the Cotton: The Southern Tenant Farmers' Union and the New Deal*. Chapel Hill: University of North Carolina Press, 1971.

Gunther, John. *The Story of TVA*. New York: Harper, 1953.

Hardin, Bill. Interview by Laura Beth Daws, July 15, 2012.

Hodge, George. Interview by Laura Beth Daws, July 16, 2012.

Hubbard, Preston J. *Origins of the TVA: The Muscle Shoals Controversy, 1920–1932*. Nashville: Vanderbilt University Press, 1961.

Huie, William Bradford. *Mud on the Stars*. Tuscaloosa: University of Alabama Press, 1996.

Jones, Jacqueline. "Federal Power, Southern Power: 1860–1940." *Journal of American History* 87, no. 4 (2001): 1392–1396.

Killingsworth, Steve. "Regional Planning on a Local Scale: The Brief Life of Pickwick Village." *West Tennessee Historical Society Papers* 51 (1997): 48–63.

Kitchens, Carl. "The Use of Eminent Domain in Land Assembly: The Case of the Tennessee Valley Authority." *Public Choice* 160, no. 3/4 (2014): 455–466.

Krutch, Charles. Interview by Charles W. Crawford, November 10, 1969.

Lazarsfeld, Paul F., Bernard Berelson, and Hazel Gaudet. *The People's Choice*. 2nd ed. New York: Columbia University Press, 1948.

Lee, Mordecai. *Congress vs. The Bureaucracy: Muzzling Agency Public Relations*. Norman: University of Oklahoma Press. 2011.

Leuchtenburg, William E. *Franklin D. Roosevelt and the New Deal, 1932–1940*. New York: Harper & Row, 1963.

Lilienthal, David E. Interview by Charles W. Crawford, February 6 and 7, 1970.

———. "Birmingham and the TVA Industrial Development Program." Speech delivered to the Birmingham Rotary Club, Birmingham, AL, October 31, 1934.

———. "Figures Show TVA's Electricity Plan Works." Speech delivered to the American Society of Mechanical Engineers, Norris, TN, June 22, 1935.

———. "Some Observations on the TVA." Speech delivered to TVA employees at First Methodist Church, Knoxville, TN, June 12, 1936.

———. "Talk to TVA Employees." Speech delivered at Chickamauga Dam, Chattanooga, TN, July 5, 1937.

———. "The Tennessee Valley Authority and Farm Electrification." Speech delivered to the National Convention of the American Farm Bureau Federation, Nashville, TN, December 12, 1934.

———. *TVA: Democracy on the March*. New York: Harper & Row, 1953.

———. *The Journals of David E. Lilienthal: Volume I: The TVA Years, 1939–1945*. New York: Harper & Row, 1964.

Martin, Thomas. *The Story of Electricity in AL since the Turn of the Century, 1900–1952*. Birmingham: Alabama Power Company, 1952.

McCarthy, Charles. Interview by Charles W. Crawford, October 30, 1969.

McCombs, Maxwell E., and Donald L. Shaw. "The Agenda-Setting Function of Mass Media." *The Public Opinion Quarterly* 36, no.2 (1972): 176–187.

McCraw, Thomas K. *TVA and the Power Fight, 1933–1939*. Philadelphia: J. B. Lipincott. 1971.

McDonald, Michael J., and John Muldowny. *TVA and the Dispossessed: The Re-*

settlement of Population in the Norris Dam Area. Knoxville: University of Tennessee Press, 1981.

Mertz, Paul E. *New Deal Policy and Southern Rural Poverty.* Baton Rouge: Louisiana State University Press, 1978.

Morgan, Arthur E. *The Making of the TVA.* Buffalo, NY: Prometheus Books, 1974.

Morgan, Harcourt A. "A National Program in the Tennessee Valley." Speech, Washington, DC, September 26, 1934.

Myers, William Starr. *The State Papers of Herbert Hoover.* New York: Doubleday, 1934.

Nichols, William I. "Teaching Grandmother How to Spin: TVA and the Private Utilities. *Harper's Magazine,* July 1936.

Nixon, Herman C. *Forty Acres and Steel Mules.* Chapel Hill: University of North Carolina Press, 1938.

Norris, George W. *Fighting Liberal: The Autobiography of George W. Norris.* New York: Collier Books, 1961.

Owen, Marguerite. *The Tennessee Valley Authority.* New York: Praeger, 1973.

Parker, Nancy Cabaniss. Interview by Laura Beth Daws, July 16, 2012.

Pritchett, C. H., and C. M. Stephenson. *Preliminary and Confidential Report: The Guntersville Area and the Proposed Coles Bend Bar Dam.* Washington, DC: TVA, November 23, 1935.

Purcell, Aaron D. *Arthur Morgan: A Progressive Vision for American Reform.* Knoxville: University of Tennessee Press, 2014.

———. *White Collar Radicals: TVA's Knoxville Fifteen, the New Deal, and the McCarthy Era.* Knoxville: University of Tennessee Press, 2009.

Report on the Economic Conditions of the South. Washington, DC: National Emergency Council, 1938.

Roberts, Charles Kenneth. *The Farm Security Administration: Rural Rehabilitation in the South.* Knoxville: University of Tennessee Press, 2015.

———. "New Deal Community-Building in the South: Subsistence Homesteads around Birmingham, AL." *The Alabama Review* 66, no. 2 (1997): 83–121.

Rogers, William Warren, Robert David Ward, Leah Rawls Atkins, and Wayne Flynt. *Alabama: The History of a Deep South State.* Tuscaloosa: University of Alabama Press, 1994.

Rothrock, Mary Utopia. Interview by Charles W. Crawford, January 16, 1970.

Schlesinger, Arthur M. *The Coming of the New Deal.* Boston: Houghton Mifflin, 1959.

Seavoy, Ronald E. *The American Peasantry: Southern Agricultural Labor and Its Legacy, 1850–1995.* Westport, CT: Greenwood Press, 1998.

Selznick, Philip. *TVA and the Grass Roots: A Study of Politics and Organization.* Berkeley: University of California Press, 1984.

Shelton, Barrett, Sr. Interview by Charles W. Crawford, June 18, 1970.

Simonson, Eugene. Interview by Laura Beth Daws, July 19, 2012.

Smith, Jean Edward. *FDR.* New York: Random House, 2008.

Swidler, Joseph. Interview by Charles W. Crawford, October 29, 1969, and December 8, 1969.

Tennessee Valley Authority Act of 1933. HR 5081. 73rd Cong., 1st sess., *Congressional Record.*

Tennessee Valley Authority (TVA). *Activities of the Reservoir Family Removal Section, Wheeler Reservoir Area.* In TVA Population Removal Records, Family Relocation Files 1934–1954, RG142 Box 59, National Archives–SERA.

———. *Agricultural-Industrial Survey of Marshall County, Alabama.* Directed by Woolrich, W. R. for the Tennessee Valley Authority, conducted by the Civil Works Administration, edited by the Industry Division of TVA. Washington, DC: TVA,1935.

———. *A History of Navigation on the Tennessee River: An Interpretation of the Economic Influence of this River System on the Tennessee Valley.* Washington, DC: TVA, 1937.

———. *A Technical Review of the Wheeler Project.* Technical Monograph no. 38. Washington, DC: TVA, June 1938.

———. *The Guntersville Project: A Comprehensive Report on the Planning, Design, Construction and Initial Operations of the Guntersville Project.* Technical Report no. 4. Knoxville, TN: TVA, 1941.

———. *Population Readjustment, Guntersville Area.* Knoxville, TN: TVA, June 1940.

———. "Population Readjustment in the Tennessee Valley." Unpublished manuscript accessed at TVA Library, Knoxville, TN.

———. *The Wheeler Project Technical Report 2.* Knoxville, TN: TVA, 1939.

———. *Wheeler Project–Final Report (First Draft), Report on Family Removal Activities.* Knoxville, TN: TVA.

TVA Board of Directors. *Report to Congress on the Unified Development of the Tennessee River System.* Knoxville, TN: TVA, March 1936.

TVA Land Acquisition Division. *Instructions to Land Buyers, Reservoir Purchase.* Knoxville, TN: TVA, June 1936.

TVA Office of Engineering, Design, and Construction, Project Histories and Reports, RG 142 Box 26, National Archives–SERA.

TVA Office of the General Manager. Information Office Correspondence Files, 1933–1946, RG142 Box 23. Atlanta: National Archives–Southeast Region (SERA).

———. Records of the Board of Directors, David E. Lilienthal General Correspondence, 1933–1946, RG 142. Atlanta: SERA.

———. *TVA Housing Development at Pickwick Dam.* Information Office Correspondence Files, 1933–1946, RG 142 Box 23. Atlanta: SERA.

TVA Population Removal Records, Family Relocation Files 1934–1954. Atlanta: SERA.

TVA Research Section, Social and Economic Division. *Preliminary and Confidential Report: The Guntersville Area and the Proposed Coles Bend Bar Dam.* Knoxville, TN: TVA, November 23, 1935.

TVA Reservoir Property Management Department, Population Readjustment Division. *Family Case Records Population Readjustment.* Knoxville, TN: TVA.

Thompson, Hazel Moore. Interview by Laura Beth Daws, June 16, 2012.

Tidwell, Luther and Beryl. Interview by Laura Beth Daws, July 16, 2012.

"Unemployment Relief Census." *Monthly Labor Review* 39 (1934): 31.

US Census Bureau. 1930 census.

———. 1940 census.

———. "Populations, Farms and Farm Property–Summary, 1850–1935," Table 603 (accessed July 1, 2014). http://www2.census.gov/prod2/statcomp/documents/1940–07.pdf.

———. "Hours and Earnings," Table 374, 1930 census.

———. "Net Monthly Bill for Specified Quantities of Electric Energy, Based on Rates as of Dec. 15 1938 and 1939, by Cities," Table 456 (accessed July 1, 2014), http://www2.census.gov/prod2/statcomp/documents/1940–05.pdf.

———. "Statistical Summary of Education," 1940 census.

———. "Farms—Number, Acreage and Value, by Color of Operator for North and West, and by Color and Tenure of Operators, for South, by States: 1930 and 1935," Table 616.

Whatley, Warren C. "Labor for the Picking: The New Deal in the South." *Journal of Economic History* 43, no. 4 (1983): 905–929.

Whitman, Wilson. *God's Valley: People and Power along the Tennessee River.* New York: The Viking Press, 1939.

Wright, Gavin. "The New Deal and the Modernization of the South." *Federal History,* 2010 (accessed 26 November 2017), http://shfg.org/shfg/wp-content/uploads/2011/01/5-Wright-design5-_Layout-1.pdf.

Index